Springer Complexity

Springer Complexity is an interdisciplinary program publishing the best research and academic-level teaching on both fundamental and applied aspects of complex systems – cutting across all traditional disciplines of the natural and life sciences, engineering, economics, medicine, neuroscience, social and computer science.

Complex Systems are systems that comprise many interacting parts with the ability to generate a new quality of macroscopic collective behavior the manifestations of which are the spontaneous formation of distinctive temporal, spatial or functional structures. Models of such systems can be successfully mapped onto quite diverse "real-life" situations like the climate, the coherent emission of light from lasers, chemical reaction-diffusion systems, biological cellular networks, the dynamics of stock markets and of the internet, earthquake statistics and prediction, freeway traffic, the human brain, or the formation of opinions in social systems, to name just some of the popular applications.

Although their scope and methodologies overlap somewhat, one can distinguish the following main concepts and tools: self-organization, nonlinear dynamics, synergetics, turbulence, dynamical systems, catastrophes, instabilities, stochastic processes, chaos, graphs and networks, cellular automata, adaptive systems, genetic algorithms and computational intelligence.

The two major book publication platforms of the Springer Complexity program are the monograph series "Understanding Complex Systems" focusing on the various applications of complexity, and the "Springer Series in Synergetics", which is devoted to the quantitative theoretical and methodological foundations. In addition to the books in these two core series, the program also incorporates individual titles ranging from textbooks to major reference works.

Editorial and Programme Advisory Board

Understanding Complex Systems

Founding Editor: J.A. Scott Kelso

Future scientific and technological developments in many fields will necessarily depend upon coming to grips with complex systems. Such systems are complex in both their composition – typically many different kinds of components interacting simultaneously and nonlinearly with each other and their environments on multiple levels – and in the rich diversity of behavior of which they are capable.

The Springer Series in Understanding Complex Systems series (UCS) promotes new strategies and paradigms for understanding and realizing applications of complex systems research in a wide variety of fields and endeavors. UCS is explicitly transdisciplinary. It has three main goals: First, to elaborate the concepts, methods and tools of complex systems at all levels of description and in all scientific fields, especially newly emerging areas within the life, social, behavioral, economic, neuro- and cognitive sciences (and derivatives thereof); second, to encourage novel applications of these ideas in various fields of engineering and computation such as robotics, nano-technology and informatics; third, to provide a single forum within which commonalities and differences in the workings of complex systems may be discerned, hence leading to deeper insight and understanding.

UCS will publish monographs, lecture notes and selected edited contributions aimed at communicating new findings to a large multidisciplinary audience.

Philippe Blanchard · Dimitri Volchenkov

Mathematical Analysis
of Urban Spatial Networks

 Springer

Philippe Blanchard
Dimitri Volchenkov
University of Bielefeld
Research Center BiBoS –
Bielefeld-Bonn-Stochastics
Universitätsstr. 25
33615 Bielefeld
Germany
blanchard@physik.uni-bielefeld.de
dima427@yahoo.com

ISBN: 978-3-540-87828-5   e-ISBN: 978-3-540-87829-2

DOI 10.1007/978-3-540-87829-2

Understanding Complex Systems ISSN: 1860-0832

Library of Congress Control Number: 2008936493

Cover design: WMXDesign GmbH

Printed on acid-free paper

9 8 7 6 5 4 3 2 1

springer.com

To our wives,
Françou and Lyudmila,
and sons,
Nicolas, Olivier, Dimitri, Andreas,
and Wolfgang.

Preface

"We shape our buildings, and afterwards our buildings shape us," said Sir Winston Churchill in his speech to the meeting in the House of Lords, October 28, 1943, requesting that the House of Commons bombed out in May 1941 be rebuilt exactly as before. Churchill believed that the configuration of space and even its scarcity in the House of Commons played a greater role in effectual parliament activity. In his view, "giving each member a desk to sit at and a lid to bang" would be unreasonable, since "the House would be mostly empty most of the time; whereas, at critical votes and moments, it would fill beyond capacity, with members spilling out into the aisles, giving a suitable sense of crowd and urgency," [Churchill].

The old Houseof Commons was rebuilt in 1950 in its original form, remaining insufficient to seat all its members.

The way you take this story depends on how you value your dwelling space – our appreciation of space is sensuous rather than intellectual and, therefore, relys on the individual culture and personality. It often remains as a persistent birthmark of the land use practice we learned from the earliest days of childhood.

In contrast to the individual valuation of space, we all share its immediate apprehension, "our embodied experience" (Kellert 1994), in view of Churchill's intuition that the influence of the built environment on humans deserves much credit.

Indeed, the space we experience depends on our bodies – it is what makes the case for near and a far, a left and a right (Merleau-Ponty 1962). On the small scale of actual human hands-on activity, the world we see is identified as the objective external world from which we can directly grasp properties of the objects of perception. A collection of empirically discovered principles concerning the familiar space in our immediate neighborhood is known as Euclidean geometry formulated in an ideal axiomatic form by Euclid circa 300 BC.

However, it was demonstrated by Hatfield (2003) that on a large scale our visual space differs from physical space and exhibits contractions in all three dimensions with increasing distance from the observer. Furthermore, the experienced features of this contraction (including the apparent convergence of lines in visual experience that are produced from physically parallel stimuli in ordinary viewing conditions) are not the same as would be the experience of a perspective projection onto a plane (Hatfield 2003).

As a matter of fact, the built environment constrains our visual space thus limiting our space perception to the immediate Euclidean vicinities and structuring a field of possible actions in that. By spatial organization of a surrounding place, we can create new rules for how the neighborhoods where people can move and meet other people face-to-face by chance are fit together on a large scale into the city.

In our book, we address these rules and show how the elementary Euclidean vicinities are combined into a global urban area network, and how the structure of the network could determine human behavior.

Cities are the largest and probably among the most complex networks created by human beings. The key purpose of built city elements (such as streets, places, and buildings) is to create the spaces and interconnections that people can use (Hillier 2004). As a rule, these elements originate through a long process of growth and gradual development spread over the different historical epochs. Each generation of city inhabitants extends and rearranges its dwelling environment adapting it according to the immediate needs, before passing it onto the next generation. In its turn, the huge inertia of the existing built environment causes chief social and economic impacts on the lives of its inhabitants. An emergent structure of the city is considered a distributed process evolving with time from innumerable local actions rather than as an object.

Studies of urban networks have a long history. In many aspects, they differ substantially from other complex networks found in the real world and call for an alternative method of analysis.

In our book, we discuss methods which may be useful for spotting the relatively isolated locations and neighborhoods, detecting urban sprawl, and illuminating the hidden community structures in complex fabric of urban area networks.

In particular, we study the compact urban patterns of two medieval German cities (the downtown of Bielefeld in Westphalia and Rothenburg ob der Tauber in Bavaria); an example of the industrial urban planning mingled together with sprawling residential neighborhoods – Neubeckum, the important railway junction in Westphalia; the webs of city canals in Venice and in Amsterdam, and the modern urban development of Manhattan, a borough of New York City planned in grid.

Although we use the methods of spectral graph theory, probability theory, and statistical physics, as should be evident from the contents, it was not our intent to develop these theories as the subject that has already been done in detail and from many points of view in the special literature. We do not give proof for most of the classical theorems referring interested readers to the special surveys. Throughout, we have tried to demonstrate how these methods, while applying in synergy to urban area networks, create a new way of looking at them.

We include as much background material as necessary and popularize it by a large scale, so that the book can be read by physicists, civil engineers, urban planners, and architects with a strong mathematical background – all those actively involved in the management of urban areas, as well as other readers interested in urban studies.

This book is targeted to bring about a more interdisciplinary approach across diverse fields of research including complex network theory, spectral graph theory, probability theory, statistical physics, and random walks on graphs, as well as sociology, wayfinding and cognitive science, urban planning, and traffic analysis.

The subsequent five chapters of this book describe the emergence of complex urban area networks, their structure and possible representations (Chap. 1). Chapters 2 and 3 review the methods of how these representations can be investigated. Chapter 4 extends these methods on the cases of directed networks and multiple interacting networks (say, the case of many transportation modes interacting with each other by means of passengers). Finally, in Chapter 5, we review the evidence of urban sprawl's impact, examine the possible redevelopments of sprawling neighborhoods, and briefly discuss other possible applications of our theory.

Humans live and act in Euclidean space which they percept visually as affine space, and which is present in them as a mental form. In another circumstance we spoke of fishes: they know nothing either of what the sea, or a lake, or a river might really be and only know fluid as if it were air around them. While in a complex environment, humans have no sensation of it, but need time to construct its "affine representation" so they can understand and store it in their spatial memory. Therefore, human behaviors in complex environments result from a long learning process and the planning of movements within them. Random walks help us to find such an "affine representation" of the environment, giving us a leap outside our Euclidean aquatic surface and opening up and granting us the sensation of new space.

Last but not least, let us emphasize that the methods we present can be applied to the analysis of any complex network.

This work had been started at the University of Bielefeld, in July 2006, while one of the authors (D.V.) had been supported by the *Alexander von Humboldt Foundation* and by the DFG-International Graduate School *Stochastic and Real-World Problems*, then continued in 2007 being supported by the *Volkswagen Foundation* in the framework of the research project *"Network formation rules, random set graphs and generalized epidemic processes."*

Many colleagues helped over the years to clarify many points throughout the book. Our thanks go to Bruno Cessac, Santo Fortunato, Jürgen Jost, Andreas Krüger, Tyll Krüger, Thomas Küchelmann, Ricardo Lima, Zhi-Ming Ma, Helge Ritter, Gabriel Ruget and Ludwig Streit.

We are further indebted to Dr. Christian Caron's competent advice and assistance in the completion of the final manuscript and our referees contributed some very useful insights. Their assistance is gratefully acknowledged.

Bielefeld   Philippe Blanchard and Dimitri Volchenkov

Contents

Chapter 1
Complex Networks of Urban Environments

The very first problem of graph theory was solved in 1736 by Leonard Euler, the invited member of the newly established Imperial Academy of Science in Saint Petersburg (Russia).

Euler had answered a question about travelling across a bridge network in the capital Prussian city of Königsberg (presently Kaliningrad, Russia) set on the Pregel River, and included two large islands which were connected to each other and the mainland by seven bridges (see Fig. 1.1).

The question was whether it was possible to visit all churches in the city by walking along a route that crosses each bridge exactly once, and return to the starting point. Euler formulated the problem of routing by abstracting the case of a particular city, by eliminating all its features except the landmasses and the bridges connecting them, by replacing each landmass with a dot, called a vertex or node, and each bridge with an arc, called an edge or link.

When discussing graphs, many intuitive ideas become mathematical and get quite natural names. A path is a sequence of vertices of a graph such that from each of its vertices there is an edge to the next vertex in the sequence. A cycle is a path such that the start vertex and end vertex are the same. A path with no repeated vertices is called a simple path, and cycle with no repeated vertices aside from the start/end vertex is a simple cycle.

Euler realized that the solution to the problem can be expressed in terms of the degrees of the nodes. The degree of a node in an undirected graph is the number of edges that are incident to it. In particular, a simple cycle of the desired form is possible if and only if there are no nodes of odd degree. Such a path is now called an Euler tour – in 1736, in Königsberg, it was not the case.

The message beyond Euler's proof was very profound: topological properties of graphs (or networks) may limit or, quite the contrary, enhance our aptitude for travel and action in them.

In the 1944 bombing, Königsberg suffered heavy damage from British air attacks and burnt for several days – two of the seven original Königsberg bridges were destroyed. Two others were later demolished by the Russian administration and replaced by a modern highway. The other three bridges remain, although only two of them are from Euler's time (one was rebuilt by the Germans in 1935) (Taylor 2000, Stallmann 2006).

Ph. Blanchard, D. Volchenkov, *Mathematical Analysis of Urban Spatial Networks,*
Understanding Complex Systems, DOI 10.1007/978-3-540-87829-2_1,
© Springer-Verlag Berlin Heidelberg 2009

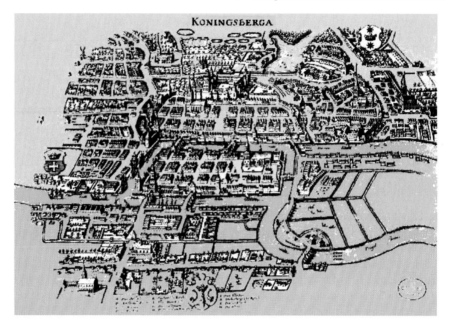

Fig. 1.1 Königsberg in 1652 by (Stich) von Merian-Erben. The picture has been taken from the Preussen-Chronik portal at $http://www.preussen-chronik.de$. The Seven Bridges of Königsberg is a famous mathematical problem inspired by the puzzle of the bridges on the Pregel. The burghers of Königsberg wondered whether it was possible to plan a walk in such a way that each bridge would be crossed once and only once

By the way, two of the nodes in the graph of Königsberg now have degree 2, and the other two have degree 4, therefore, an Eulerian tour is possible in the city today, although no services are held in the churches.

Various practical studies related to the city and intercity routing problems have been a permanent issue of inspiration for graph theory. For instance, one can mention the travelling salesman problem, in which the cheapest round-trip route is searched such that the salesman visits each city exactly once and then returns to the starting city (Dantzig et al. 1954). The shortest path searching algorithms and minimum spanning trees originated in 1926 for the purpose of efficient electrical coverage of Bohemia (Nesetril et al. 2000). Routing studies underwent a rapid progression due to the development of probability theory and the implication of random walks in particular.

In mathematics and physics, a random walk is a formalization of the intuitive idea of taking successive steps, each in a random direction. The term "random walk" was originally proposed by K. Pearson in 1905 in his letter to the "Nature" journal devoted to a simple model describing a mosquito infestation in a forest: at each time step, a single mosquito moves a fixed length, at a randomly chosen angle (see Hughes 1996).

Lagrange was probably the first scientist who investigated a simple dynamical process (diffusion) in order to study the properties of a graph (Lagrange 1867). He calculated the spectrum of the Laplace operator defined on a chain (a linear graph) of N nodes in order to study the discretization of the acoustic equations. Today, it is well-known that random walks could be used in order to investigate and characterize how effectively the nodes and edges of large networks can be covered by different strategies (see Tadic 2002, Yang 2005, Costa et al. 2007 and many others). The simplest model is called a "drunkard's walk."

Imagine a drunkard wandering along the streets of an ideally "gridded" city and choosing one of the four possible routes (including the one he came from) with equal probability. It is known from studies of random walks on a planar lattice that, following such a strategy, the drunkard was almost surely get back to his home from the bar sooner or later (Hughes 1996). The trajectory of the drunkard in this case is just the sequence of street junctions he passed through, regardless of when the walk arrived at them.

However, if the city is no longer a perfect square grid, but the drunkard is still using his random strategy while choosing the direction of movement, the probability that he chooses a street changes from one junction to another and, therefore, in a probabilistic sense, his forthcoming trajectory depends upon the previous steps. As a matter of fact, the trajectory of the drunkard is a fingerprint of the graph topology. By analyzing the statistical properties of a large number of random walk trajectories on the given graph, we can obtain information about some of its important topological properties.

Being motivated by many practical applications, the random walks defined on graphs and the tightly related diffusion processes have been studied in detail. In economics, the random walk hypothesis is used to model share prices and other factors (Keane 1983). In population genetics, the random walk describes the statistical properties of genetic drift (Cavalli-Sforza 2000). During WWII a random walk was used to model the distance that an escaped prisoner of war would travel in a given time. In psychology, random walks accurately explain the relation between the time needed to make a decision and the probability that a certain decision will be made. Random walk can be used to sample from a state space which is unknown or very large, for example to pick a random page of the internet or, for research of working conditions of a random illegal worker (Hughes 1996). Random walks are often used in order to reach the "obscure" parts of large sets and estimate the probable access times to them (Lovasz 1993). Sampling by random walk was motivated by important algorithmic applications to computer science (see Deyer et al. 1986, Diaconis 1988, Jerrum et al. 1989). There are a number of other processes that can be defined on a graph describing various types of diffusion of a large number of random walkers moving on a network at discrete time steps (Bilke et al. 2001).

The attractiveness of random walks and diffusion processes defined on the undirected non-bipartite graphs is due to the fact that the distribution of the current node after t steps tends to a well-defined stationary distribution π which is uniform if the graph is regular. In contrast to them, there could be in general no any stationary distribution for directed graphs (Lovasz et al. 1995).

However, before we can apply graph theoretic tools to the urban networks, we must parse the geometry of an urban space and translate it into a pattern that supports the type of analysis to be performed. Despite the long tradition of research articulating urban area form using graph-theoretic principles, this step is not as easy as it may appear.

1.1 Paradigm of a City

A city has often been compared with a biological entity (Miller 1978) – a single organism covering the entire landscape surface and showing signs of a vast intelligence. It is well-known that many physiological characteristics of biological organisms scale with the mass of their bodies M (Savage et al. 2006). For example, the power Pw required to sustain a living organism takes the shape of a straight line on the logarithmic scale (West et al. 1998):

$$Pw \propto M^{3/4}.$$

It is remarkable that all important demographic and socioeconomic urban indicators such as consumption of energy and resources, production of artifacts, waste, and even greenhouse gas emissions alike scale with the size of a city (Bettencourt et al. 2007). The pace of life in cities also increases with the size of population: wages, income, growth domestic product, bank deposits, as well as rates of invention, measured by the number of new patents and employment in creative sectors scale superlinearly over different years and nations (Florida 2004).

However, while it is suggested that humans study cities, the opposite seems to be the case, where cities examine and reveal the hidden essences of men by integrating and segregating them at the same time. A city that is organic, sentient, and powerful allows us to hover over it in an attempt to fathom some of its mysteries.

Will the human mind ever understand this form of life?

1.1.1 Cities and Humans

A belief in the influence of the built environment on humans was common in architectural and urban thinking for centuries. Cities generate more interactions with more people than rural areas because they are central places of trade that benefit those who live there.

People moved to cities because they intuitively perceived the advantages of urban life. City residence brought freedom from customary rural obligations to lord, community, or state and converted a compact space pattern into a pattern of relationships by constraining mutual proximity between people.

A long time ago, city inhabitants began to take on specialized occupations, where trade, food storage and power were centralized. Benefits included reduced transport costs, exchange of ideas, and sharing of natural resources.

A key result of urbanization has been an increased division of labor and the growth of occupations geared toward innovation and wealth creation responsible for the large diversity of human activity and organization (Durkheim 1964). The impact of urban landscapes on the formation of social relations studied in the fields of ethnography, sociology, and anthropology suggests it is a crucial factor in the technological, socioeconomic, and cultural development (Low et al. 2003).

The increasing development density has the advantage of making mass transport systems, district heating and other community facilities (schools, health centers, etc.) more viable (Kates et al. 2003). At the same time, cities are the main sources of pollution, crime, and health problems resulting from contaminated water and air, and communicable diseases.

Life in a city changes our nature, our perceptions and emotions, and the way we would relate to others. Massive urbanization, together with a dramatic increase in life expectancy, gives rise to a phenomenon where innovations occur on time scales that are much shorter than individual life spans and shrinking further as urban population increases, in contrast to what is observed in biology. Moreover, major innovation cycles must be generated at a continually accelerating rate to sustain growth and avoid stagnation or collapse of cities (Bettencourt et al. 2007).

Spatial organization of a place has an extremely important effect on the way people move through spaces and meet other people by chance (Hillier et al. 1984). Compact neighborhoods can foster casual social interactions among neighbors, while creating barriers to interaction with people outside a neighborhood. Spatial configuration promotes people's encounters as well as making it possible for them to avoid each other, shaping social patterns (Ortega-Andeane et al. 2005). Segregation is a part of this complex phenomenon. When a city appears to be profoundly indifferent to humanity, and life in that city becomes a symbol of remoteness and loneliness.

The phenomenon of clustering of minorities, especially that newly arrived immigrants, is well documented since the work of Wirth 1928 (the reference appears in Vaughan 2005a). Clustering is considering to be beneficial for mutual support and for the sustenance of cultural and religious activities. At the same time, clustering and the subsequent physical segregation of minority groups would cause their economic marginalization. The study of London's change over 100 years performed by Vaughan et al. (2005b) has indicated that the creation of poverty areas is a spatial process; by looking at the distribution of poverty on the street it is possible to find a relationship between spatial segregation and poverty. The patterns of mortality in London studied over the past century by Orford et al. (2002) show that the areas of persistence of poverty cannot be explained other than by an underlying spatial effect.

Immigrant quarters have historically been located in the poorest districts of cities. The spatial analysis of the immigrant quarters reported in Vaughan (2005a) shows that they were significantly more segregated from the neighboring areas, in particular, there were less streets turning away from the quarters to the city centers than in the other inner-city areas which were usually socially barricaded by the railway and industries (Williams 1985). The poorer classes have been found to be dispersed over the spatially segregated pockets of streets formed by the interruption of the city grid due to railway lines and large industrial buildings, and the urban structure itself encourages the economic conditions for segregation.

The spatial distribution of poverty had been taken into account in the United States while formulating the mortgage terms and policies at the height of the depression in the early 1930s. With the National Housing Act of 1934 which established the Federal Housing Administration, the "residential security maps" had been created for 239 cities in order to indicate the level of security for real estate investments in each surveyed city – the lending institutions were extremely reluctant to make loans for housing. In these maps many minority neighborhoods in cities were not eligible to receive housing loans at all. The practice of marking a red line on a map to delineate the area where banks would not invest is known as redlining. The redlining together with a policy of withdrawing essential city services (such as police patrols, garbage removal, street repairs, and fire services) from neighborhoods suffering from urban decay, crime and poverty have resulted in even large increase in residential segregation, urban decay and city shrinking (Thabit 2003). Shrinking cities have become a global phenomenon: the number of them has increased faster in the last 50 years than the number of expanding ones.

Places that are unpleasant and alienating will cause people to avoid them to the best of their ability and, therefore, a city decline can be initiated by a spatial process of physical segregation. Reducing movement in a spatial pattern is crucial for the decline of new housing areas. Spatial structures creating a local situation in which there is no relation between movements inside the compact neighborhood and outside it, and the lack of natural space occupancy becomes associated with the social misuse of the structurally abandoned spaces (Hillier 2004).

It is well-known that the urban layout has an effect on the spatial distribution of crime (UK Home Office). Furthermore, different types of crime are associated with different levels of land use and social characteristics (Dunn 1980). Personal attack crimes occur in lower class neighborhoods, while property crimes occur in neighborhoods that are accessible or close to land uses, or in neighborhoods with higher percentages of underemployed or single residents. Robberies and burglaries share monetary gain objectives and are more likely to occur in middle- and upper-class neighborhoods (Rengert 1980). Crime seems to be highest where the urban grid is most broken up (in effect creating most local segregation), and lowest where the lines are longest and most integrated (Hillier 2004).

1.1.2 Facing the Challenges of Urbanization

Multiple increases in urban population that had occurred in Europe at the beginning of the 20th Century have been among the decisive factors that changed the world. Urban agglomerations had suffered from comorbid problems such as widespread poverty, high unemployment, and rapid changes in the racial composition of neighborhoods. Riots and social revolutions have occurred in urban places in many European countries, partly in response to deteriorated conditions of urban decay and fostered political regimes affecting immigrants and certain population groups de facto alleviating the burden of the haphazard urbanization by increasing its deadly price. Tens of millions of people emigrated from Europe, but many more of them died of starvation and epidemic diseases, or became victims of wars and political repressions.

Today, according to the United Nations Population Fund report (UN 2007), the accumulated urban growth in the developing world will be duplicated in a single generation. By the end of 2008, for the first time in human history, more than half of the population (hit 3.3 billion people) will be living in urban areas. Thirty years from now is expected to be a decisive period for humans facing the challenge of disastrous urbanization – about 5 billion people will dwell in cities, and 6.4 billion will be living in cities by the year 2050, said Reuters (2008).

Although the intense process of urbanization is proof of economic dynamism, clogged roads, dirty air, and deteriorating neighborhoods are fueling a backlash against urbanization that, nevertheless, cannot be stopped. According to the U.S. Census Bureau (U.S.Census 2006), urbanized areas in the United States sprawled out over an additional 41,000 km^2 over the last 20 years – covering an area equivalent to the entire territory of Switzerland with asphalt, buildings and subdivisions of suburbia. Between 1990–2000 the growth of urban areas and associated infrastructure throughout Europe consumed more than 8,000 km^2, an area equal to the entire territory of the Luxembourg (EEA 2006). City development planners will face great challenges in preventing cities from unlimited expansion.

The urban design decisions made today based on the U.S. car-centered model, in cities of the developing world where car use is still low, will have an enormous impact on global warming in the decades ahead – carbon dioxide from industrial and automobile emissions is a suspected cause of global warming.

Modern cities are known for creating their own micro-climates. Effects of surface warming due to urbanization on subsurface thermal regime were found in many cities over the world. Heavily urban areas where hard surfaces absorb the sun's energy, heat up, and reradiate that heat to the ambient air, resulting in the appearance of urban heat islands, cause significant downstream weather effects.

Urbanization deepens global warming; although the effect of global warming is estimated to be 0.5 degree centigrade during the last 100 years, the effects of urbanization and global warming on the subsurface environment were estimated by Taniguchi et al. (2003) to be 2.5, 2.0 and 1.5 degree centigrade in Tokyo, Osaka and Nagoya, respectively.

Sea level rise caused by thermal expansion of water and the melting of glaciers and ice sheets will have potentially huge consequences since over 60 percent of the population worldwide lives within 100 km of the coast (GEO-4 2007).

Unsustainable pressure on resources causes the increasing loss of fertile lands through degradation, and the dwindling amount of fresh water and food would trigger conflicts and result in mass migrations. Migrations induce a dislocation and disconnection between the population and their ability to undertake traditional land use (Fisher 2008). Major metropolitan areas and the intensively growing urban agglomerations attract large numbers of immigrants with limited skills. Many of them will become a burden on the state, and perhaps become involved in criminal activity.

The poor are urbanizing faster than the population as a whole (Ravallion 2007). Global poverty is quickly becoming a primarily urban phenomenon in the developing world; among those living on no more than $1 a day, the proportion found in urban areas rose from 19 percent to 24 percent between 1993 and 2002. About 70 percent of 2 billion new urban settlers in the next 30 years will live in slums, in addition to the 1 billion already there. The fastest urbanization of poverty occurred in Latin America, where the majority of the poor now live in urban areas.

Faults in urban planning, poverty, redlining, immigration restrictions and clustering of minorities dispersed over the spatially isolated pockets of streets trigger urban decay, a process by which a city falls into a state of disrepair. The speed and scale of urban growth require urgent global actions to help cities prepare for growth and to prevent them from becoming the future epicenters of poverty and human suffering.

We have used the population data reported in (UN 2007) in order to plot the total population in 221 countries (Serbia and Montenegro have been accounted as a single state) vs. their urban population from 1950 (violet points) to 2005 (red points) by five years (see Fig. 1.2). It is remarkable that the scattering plot consists of 221 visible strokes, and each one traces a national road a country follows toward urbanization in the last 55 years. In most of the countries, the population living in urban areas grows much faster than the total population, but the rate of urbanization varies from country to country. The straight line,

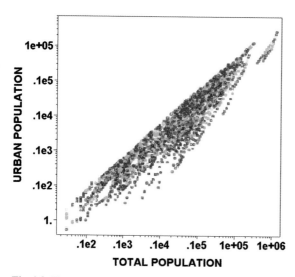

Fig. 1.2 The worldwide urbanization trend (thousands of people): the total population in 221 countries (Serbia and Montenegro have been accounted as a single state) vs. their urban population from 1950 (*violet points*) to 2005 (*red points*) by five years, as reported in (UN 2007)

$$\{ \text{Total population} \} = \{ \text{Urban population} \}, \qquad (1.1)$$

clearly seen on the plot represents the horizon of full urbanization already achieved by many countries by 2005.

Interestingly, the groups of sequential data points apparently form the segments of straight lines in the logarithmic scale providing evidence in favor of the strong persistency of the national land use systems maintaining a balance between economic development and environmental quality – town and country planning is the part of that with an essentially far-reaching influence on society. The layout of streets and squares, the allocation of parks and other open spaces, and in particular the way how they are connected to each other within a city district determine the prosperity and lives of many not only for the present, but for generations to come.

An inexorable worldwide trend toward urbanization presents an urgent challenge to develop a quantitative theory of urban organization and its sustainable development. The density of urban environments relates directly to the need to travel within them. Good quality of itineraries and transport connections is a necessary condition for continually accelerating the rate of innovations necessary to avoid stagnation and collapse of a city.

1.1.3 The Dramatis Personæ. How Should a City Look?

The joint use of scarce space creates life in cities which is driven largely by microeconomic factors which tend to give cities similar structures. The emergent street configuration creates differential patterns of occupancy, whereby some streets become, over time, more highly used than others (Iida et al. 2005). At the same time, a background residential space process driven primarily by cultural factors tends to make cities different from each other, so that the emergent urban grid pattern forms a network of interconnected open spaces, being a historical record of a city creating process driven by human activity and containing traces of society and history (Hanson 1989).

Surprisingly, there is no accepted standard international definition of a city. The term may be used either for a town possessing city status, for an urban locality exceeding an arbitrary population size, or for a town dominating other towns with particular regional economic or administrative significance. In most parts of the world, cities are generally substantial and nearly always have an urban core, but in the United States many incorporated areas which have a very modest population, or a suburban or even mostly rural character, are also designated as cities.

Modern city planning has seen many different schemes for how a city should look.

The most common pattern favored by the Ancient Greeks and the Romans, used for thousands of years in China, established in the south of France by various rulers, and being almost a rule in the British colonies of North America is the grid. In most

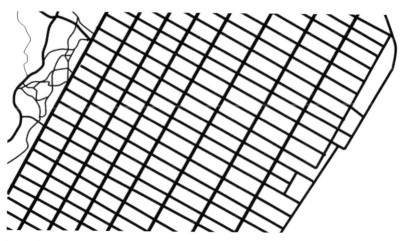

Fig. 1.3 The route scheme of the Upper East Side, a 1.8 square mile (4.7 km^2) neighborhood in the borough of Manhattan in New York City, USA, between Central Park and the East River

cities of the world that did not develop and expand over a long period of time, streets are traditionally laid out on a grid plan.

Manhattan, a borough of New York City, is a paradigmatic example (see Fig. 1.3), with the standard city block, the smallest area that is surrounded by streets, of about 80 meters by 271 m. It is coextensive with New York County, the most densely populated county in the United States, and is the sixth most populous city in the country.

Cities founded after the advent of the automobile and planned accordingly tend to have expansive boulevards impractical to navigate on foot. However, unlike many settlements in North America, New Amsterdam (Manhattan) had not been developed in grids from the beginning. In 1811 three-man commission had slapped a mesh of rectangles over Manhattan, from 14th Street on up to the island's remote wooded heights, motivated by economic efficiency ("right-angled houses ... are the most cheap to build") as well as political acceptance (as "a democratic alternative to the royalist avenues of Baroque European cities"). As the city grew into its new pattern, preexisting lanes and paths that violated the grid were blocked up, and the scattered buildings that lined them torn down. Only Broadway, the old Indian trail that angled across the island, survived (Brookhiser 2001).

Older cities appear to be mingled together, without a rigorous plan. This quality is a legacy of earlier unplanned or organic development, and is often perceived by today's tourists to be picturesque. They usually have a hub, or a focus of several directional lines, or spokes which link center to edge, and sometimes there is a rim of edge lines. Most of the trading centers are at the city's center, while the areas outside of the center are the more residential ones.

The spatial structure of organic cities was shaped in response to the socioeconomic activities maximizing ease of navigation in the areas, which are most likely

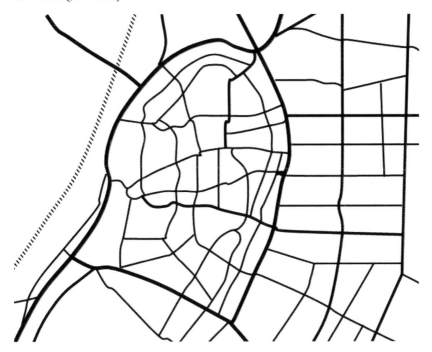

Fig. 1.4 The route scheme of the "hidden" city of Bielefeld, North Rhine-Westphalia (Germany) is an example of an organic city

to be visited by different people from inside and outside, but minimizes the same when it is undesired (Hillier 2005).

The city of Bielefeld (see Fig. 1.4), founded in 1214 by Count Hermann IV of Ravensberg to guard a pass crossing the Teutoburg Forest, represents a featured example of an organic city.

While public spaces bring people together, maximizing the reach of them and movement through them, the guard functions delegated to the city many centuries ago had sought to structure relations between inhabitants and strangers in the opposite way. A lot of people who pass through Bielefeld every day reside on the highly important passage between the region of Ruhr and Berlin, with one of the voluminous Germany highways and the high-speed railway; however it is apparent that most Germans do not have a clear image of the city in their heads. In spite of all the efforts by the city council to subsidize development and publicity for Bielefeld, it has a solid reputation for obscurity seldom found in a city its size (Bielefeld is the biggest city within the region of Eastern Westphalia). The common opinion on the "hidden" city of Bielefeld is perfectly characterized by the "Bielefeld conspiracy theory" (Die Bielefeld-Verschwörung), a sustained satirical story popular among German Internet users from May 1994 (Held). It says that the city of Bielefeld does not actually exist and is merely an alien base.

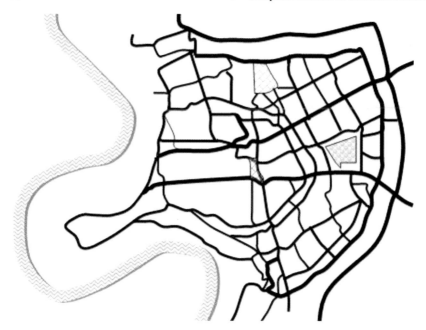

Fig. 1.5 The route scheme for Rothenburg ob der Tauber, Bavaria (Germany)

Other organic cities may show a radial structure in which main roads converge to a central point, often the effect of successive growth over time with concentric traces of town walls (clearly visible on the satellite image of the medieval Bavarian city, Rothenburg ob der Tauber Fig. 1.5) and citadels usually recently supplemented by ringroads that take traffic around the edge of a town.

Rothenburg had been founded between 960 and 970 AD, but its elevation to a free empire city occurred between 1170 and 1240. After the 30-year war (1618–1648), its development was practically quiet and the city became meaningless. Yet before World War I, Rothenburg became a popular tourist center attracting voyagers from the United Kingdom and France. However, the obvious legibility of the city brought harm to it during World War II. Being of no military importance Rothenburg was used as a replacement target (Rothenburg) and was strongly damaged by allied bomber attacks.

The central diamond within a walled city was thought to be a good design for defense. Many Dutch cities have been structured this way: a central square surrounded by concentric canals. The city of Amsterdam (see Fig. 1.6) is located on the banks of the rivers Amstel and Schinkel, and the bay IJ. It was founded in the late 12th Century as a small fishing village, but the concentric canals were largely built during the Dutch Golden Age, in the 17th Century. Amsterdam is famous for its canals, grachten. The principal canals are three similar waterways, with their ends resting on the IJ, extending in the form of crescents nearly parallel to each other and to the outer canal. Each of these canals marks the line of the city walls and moat at

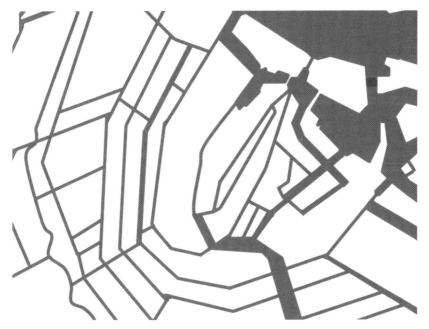

Fig. 1.6 The scheme of corals in Amsterdam, the capital city of the Netherlands

different periods. Lesser canals intersect the others radially, dividing the city into a number of islands.

Cities founded and developed in the areas bounded by natural geographical limitations (for example, on tiny islands) form a special morphological class – their structures bare the multiple fingerprints of the physical landscape.

The original population of Venice (see Fig. 1.7) was comprised of refugees from Roman cities who were fleeing successive waves of barbarian invasions (Morris 1993). From the 9th to the 12th Centuries, Venice developed into a city state. During the late 13th Century, Venice was the most prosperous city in all of Europe, dominating Mediterranean commerce. During the last millennium, the political and economical status of the city were changing, and the network of city canals was gradually redeveloping from the 9th to the early 20th Centuries. Venice is one of the few cities in the world with no cars. A rail station and parking garage are located at the edge of the city, but all travel within the city is by foot or by boat.

The historical period marking the introduction of mass production, improved transportation, and applications of technical innovations such as in the chemical industry, in canal and railway transport was accompanied by social and political changes. The rural landscapes and classical homes of the gentry were replaced by the new industrial landscapes with the identity rooted in economic production.

The small town of Neubeckum (see Fig. 1.8) is an example of an industrial place. It was founded in 1899 as the railway station of the city of Beckum on the Cologne – Minden railroad. Neubeckum has been developed as a regional railway junction and an industrial center.

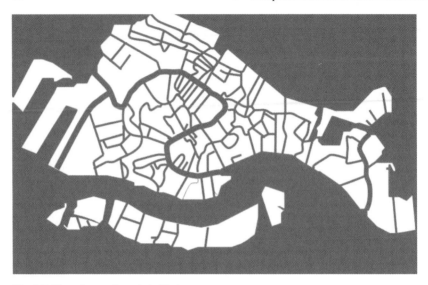

Fig. 1.7 The scheme of corals in Venice

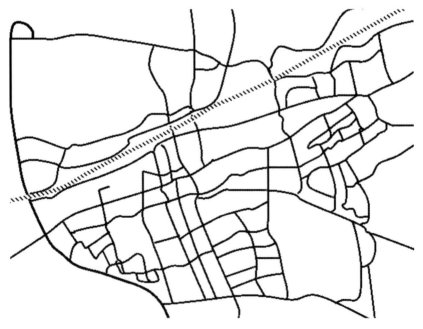

Fig. 1.8 The route scheme of the town of Neubeckum, North Rhine-Westphalia (Germany)

The scarcity of physical space is among the most important factors determining the structure of compact urban patterns. Sometimes, the historic downtowns of ancient cities can be considered as the compact urban patterns. Downtowns were the primary location of retail, business, entertainment, government, and education, but they also included residential uses. Therefore, downtowns are more densely developed than the city neighborhoods that surround them.

Some features of the compact urban patterns mentioned in the present section are given in Table 1.1 where N denotes the total number of places of motion (streets, squares, or canals) in a city, and M indicates the number of crossroads and junctions. The distance between two places of motion, A and B, is the length of the shortest path connecting them, i.e., the minimal number of other places one should cross while travelling from A to B, or vice versa. The diameter $\mathfrak{D}$ of a city graph is the distance between the two places of motion which are furthest from each other. The total number of different paths $\mathscr{P}$ is given in the last column of Table 1.1.

1.1.4 Cities Size Distribution and Zipf's Law

If we calculate the natural logarithm of the city rank in many countries and of the size of cities (measured in terms of its population) and then plot the resulting data in a diagram, we obtain a remarkable linear pattern where the slope of the line equals -1 (or $+1$, if cities have been ranked in the ascending order). In terms of the distribution, this means that the probability that the population size of a city is N

$$\Pr[\text{Population} > N] \longrightarrow N^{\zeta}. \tag{1.2}$$

The conjecture on the city population distribution can be closely approximated by a power law with $\zeta \simeq 1$ made by Auerbach (1913) (see Gabaix et al. 2004) the first explanation for that had been proposed by Zipf (1949).

The empirical validity of Zipf's Law for cities has been recently examinal using new data on the city populations from 73 countries by two different estimation methods by Soo (2002). The use of various estimators justifies the validity of Zipf's Law (1.105 for cities, but 0.854 for urban agglomerations) for 20 to 43 of the 73 investigated countries. It has been suggested that variations in the value of Zipf's exponent are better explained by political economy variables than by economic geography variables. Other complete empirical international comparative studies had

Table 1.1 Some features of the compact urban patterns we study in our book

Urban Pattern	N	M	$\mathfrak{D}$	$\mathscr{P}$
Rothenburg ob d.T.	50	115	5	10,861
Bielefeld (downtown)	50	142	6	10,938
Amsterdam (canals)	57	200	7	15,960
Neubeckum	65	254	7	18,432
Venice (canals)	96	196	5	34,762
Manhattan	355	3,543	5	415,260

been performed by Rosen et al. (1980), (the average Zipf's exponent over 44 countries is 1.13, with a standard deviation 0.19) and by Brakman et al. (1999, 2001) (the average Zipf's exponent over 42 countries is 1.13, with a standard deviation 0.19).

The main problem of all statistical studies devoted to urban size distributions is that there is no universally accepted definition of a city for statistical purposes (Gabaix et al. 2004). Differences in population data between cities and metropolitan statistical areas strongly vary from country to country, therefore making international comparisons tricky. It is clear that the exponent ζ in (1.2) is sensitive to the choice of lower population cutoff size above which a rank is assigned to a city. The Statistical Abstract of the United States lists all agglomerations larger than 250,000 inhabitants, but for a lower cutoff, the exponent ζ is typically lower (Gabaix et al. 2004).

Possible economic explanations for the rank-size distributions of the human settlements relies primarily on an interplay of transport costs, positive and negative economic feedbacks, and productivity differences (see, for instance (Brakman et al. 2001, Gabaix et al. 2004) and references therein). Another strand of research based on relatively simple stochastic models of settlement formation and growth is exemplified by Simon (1955) and Reed (2002). The classical preferential attachment approach of Simon (1955), reformulated later by Barabasi et al. (1999), has been proposed as a model of city growth in Andersson et al. (2005).

In Gabaix (1999), Gabaix et al. (2004), the Zipf's Law has been derived from the assumption that the city growth rates are independent of size: the growth process is essentially the same at all scales, so that the final distribution is scale-invariant (Gibrat's Law).

Likely the most accurate regularity in economics and in the social science, Zipf's Law constitutes a minimum requirement of admissibility for models of urban creation and growth (Gabaix 1999). In accordance with the Central Place Theory proposed by Christaller (1966), some goods and services produced in a city can be related to the population size, while others are not. Large cities, with a population above an established threshold, offer a variety of commodities that are attractive to all inhabitants in the national urban system. Cities at the second level offer a limited set of commodities and, therefore, are characterized by a smaller basin of attraction. The hierarchical structure of cities is established when the number of inhabitants in cities is balanced by their basins of attraction (Bretagnolle et al. 2006), due to asymmetric exchanges: inhabitants of small towns purchase goods from larger cities with the large scale in production, but the inverse does not happen. Recently, it has been demonstrated by Semboloni (2008) that the power-law city size distribution satisfies the balance between the offer of the city and the demand of its basin of attraction, and that the exponent in Zipf's Law corresponds to the multiplier linking the population of the central city to the population of its basin of attraction.

Finally, formal explanations relating the rank order statistics like the Zipf Law to the generalized Benford Law have been suggested in Pietronero et al. (2001). The distributions of first digits in a numberical series obtained from very different origins show a marked asymmetry in favor of small digits that falls under the name of Benford's Law. The first three integers (1–3) alone have a global frequency of 60 percent while the other six values (4–9) appear only in 40 percent of the cases. The first observation of this property traces back to Newcomb (1881), but a more precise account was given by Benford (1938) and later by Richards (1982). In Pietronero et al. (2001), it has been demonstrated that Benford's Law can be naturally explained in terms of the dynamics governed by multiplicative stochastic processes. In particular, the general Benford power-law distribution $\Pr(N) \propto N^{-\alpha}$ leads to Zipf's Law with exponent $1/(1 - \alpha)$ resulting from the ranking of numbers $\mathcal{N}$ extracted from $\Pr(N)$ if the variable N is bounded by some $N_{\max}$.

Understanding how systems with many interacting degrees of freedom can spontaneously organize into scale invariant states is of increasing interest to many fields of science. The robustness of Zipf's Law fosters additional research with enriched theories of urban growth and development.

1.1.5 European Cities: Between Past and Future

The process of urbanization in Europe has evolved as a clear cycle of change from urbanization to suburbanization to deurbanization, and to reurbanization.

The growth of modern industry from the late 18th Century led to massive urbanization, first in Europe and then in other regions. Huge numbers of migrants fled from rural communities into urban areas as new opportunities for employment in the manufacturing sector arose. Industrial manufacturing was largely responsible for the population boom cities experienced during this time period. By the late 18th Century, London had become the largest city in the world with a population of over one million.

When the population of cities increased dramatically during the late 19th and first part of the 20th Centuries the infrastructure that was in place was clearly inadequate and this discordance stimulated social and political cataclysms in many European countries. During 1950–2000, the urbanization rate increased from 64 percent to 80 percent in the United States, and from 50 percent to 71 percent in Europe. Today, approximately 75 percent of the European population live in urban areas, while still enjoying access to extensive natural landscapes (UN 2007).

By the end of 2008 for the first time in human history, one-half of the world's population will be living in cities (Reuters 2008). Though Europe will continue to lag well behind the urbanization seen elsewhere, the urban future of Europe is a matter of great concern. Eighty to 90 percents of Europeans will be living in urban areas by 2020, and tremendous changes in land use are on the horizon. Today, more than one-quarter of the EU territory has already been directly affected by urban land use, and the various demands for land in and around cities will become increasingly acute in the near future.

Recent investigations in the dynamics of the European urban network performed in the framework of the European program "Time-Geographical approaches to Emergence and Sustainable Societies" have shown that, while Europe remains one of the world's most desirable and healthy places to live despite the commercial success and attractive potentials European cities will not grow as fast in the coming century as they did in the last one (TIGrESS 2006). The primary reasons for that are, first, the decrease in fertility of the aging European population and, second, that the rural-urban migration in Europe has already reached its limits. It is clear that the demographic dynamics of Europe, and in particular that of its cities, will depend on the migration from outside Europe.

With the low demographic settings, by 2025 the European urban population will stabilize to around 460 million inhabitants, with the high demographic scenario whereas it will increase up to 600 million. In the most pessimistic case for the growth rates for Eastern Europe, a clear negative trend for urban population can be expected in all geographical areas, more accentuated for Southern and Eastern Europe. In the most optimistic demographic scenario, a slightly positive growth rate for Western Europe can be predicted, while it is negative for Eastern and Southern Europe.

Although historically the growth of cities was fundamentally linked to increasing population, more recent urbanization developments such as urban sprawl, low-density expansions of large urban areas, are no longer tied to population growth (EEA 2006). Instead, individual housing preferences, increased mobility, commercial investment decisions, and certain land use polices drive the development and growth of urban areas in modern Europe.

1.2 Maps of Space and Urban Environments

Maps provide us with the representations of urban areas that facilitate our perception and navigation of the city.

Planar graphs have long been regarded as the basic structures for representing environments where topological relations between components are firmly embedded into Euclidean space. The widespread use of graph theoretic analysis in geographic science had been reviewed in Haggett et al. (1967), establishing it as central to spatial analysis of urban environments. The basic graph theory methods had been applied to the measurements of transportation networks by Kansky (1963).

1.2.1 Object-Based Representations of Urban Environments. Primary Graphs

Any graph representation of the spatial network naturally arises from the categorization process, when we abstract the system of city spaces by eliminating all but one of its features, and by grouping places that share a common attribute by classes or categories.

There is a long tradition of research articulating urban environment form using an object-based paradigm, in which the dynamics of an urban pattern come from the landmasses, the physical aggregates of buildings delivering place for people and their activity. Under the object-based approach it is suggested a city is made up by interactions between the different components of urban environments (marked by nodes of a planar graph) measured along streets and other linear transport routes (considered as edges). The usual city plans (see Figs. 1.3, 1.4, 1.5, 1.6, 1.7, 1.8) are the examples.

Probably, the tradition of representing of urban environments by primary graphs was originated from the famous paper of L. Euler on the seven bridges of Königsberg (Prussia) (Alexanderson 2006). In our book, we call these planar graph representations of urban environments primary graphs.

In the primary graph representations, identifiable urban elements with a certain mass (residential population, building stock, business activity, and so on) are associated with locations in the Euclidean plane defined as nodes $V = \{1 \ldots N\}$, whose relationships to one another are usually based on Euclidean geometry providing spatial objects with precise coordinates along their edges and outlines. The value of links between the nodes $i \in V$ and $j \in V$ of the primary graph can be either binary, $\{0, 1\}$ – with the value 1 if i is connected to j, $i \sim j$, and 0 otherwise, or proportionate to the Euclidean distance between the nodes (expressing the connection and maintenance costs), or equal to some weight $w_{ij} \geq 0$ quantifying the dynamics coming from the flows between the discrete urban zones, i and j, induced by the "attraction" between nodes. These flows are in some sense proportional to mass and inverse proportional to distance (the recent gravity model of Korean highways (Jung et al. 2008 is an example), and the whole system works against a neutral background of metric space.

Historically, the Newtonian models developed on the base of the primary graph representation of urban environments are traditionally used as a scientific support to the planning policy of cities, but in spite of their considerable successes they have never evolved into a compelling theory of the city (Hillier 2008).

1.2.2 Cognitive Maps of Space in the Brain Network

The space we experience was conceived by Euclid of Alexandria and fitted into a comprehensive deductive and logical system of geometry. For over 2000 years, five of Euclid's axioms seemed to be self-evident statements about physical reality, so that any theorem proved from them was deemed true in an absolute sense. It was taken for granted that Euclidean geometry describes physical space, and now it is considered as a good approximation to the properties of physical space, at least if the gravitational field is not too strong.

Why does Euclidean geometry appear natural for representing the properties of space?

It is important to mention that humans are not the only creatures that perceive physical space as Euclidean. Spatial locations are a functionally important dimension in the lives of most animals. An ability to encode and remember the locations of the sources of food or homes is often critical to their survival. The results of Wilkie (1989) suggest that birds (pigeons) encode space in a Euclidean way –

i.e., for the pigeon the psychological distance between locations corresponds to Euclidean geometry distance.

The series of experiments performed earlier on rats by Cheng (1986) and Margules et al. (1988), as well as later on infants by Hermer et al. (1994) convincingly demonstrated that the brain mechanisms that subserve the navigational strategies should be somehow encapsulated in a geometric neuromodule dedicated to spatial localization and navigation in Euclidean space which is conserved across species (O'Keefe 1994). The multiple evidence that humans, apes, some birds and some small mammals appear to behave as if they have internal representations that guide way finding processes in a map-like manner have been summarized by Golledge (1999).

Navigation in mammals depends on a distributed, modularly organized brain network that computes and represents positional and directional information. Spatially coded cells have been found in several different brain regions, including parietal lobes (Andersen et al. 1985), prefrontal cortex (Funahashi et al. 1989), and superior colliculus (Mays et al. 1980) – they code the locations of objects within local coordinate frameworks centered on the retina, head or body. In contrast, cells in the hippocampus appear to code for location in an environmentally centered framework (O'Keefe 1976) becoming selectively active when the rodent visits a particular place in the environment. It has been established that hippocampal place cells are influenced by experience and may form a distributed map-like mnemonic representation of the spatial environment that the animal can use for efficient navigation (O'Keefe et al. 1978), spatial memory, object recognition memory, and for relating and combining information from multiple sources, as in learning (Broadbent et al. 2004). Head direction cells form another key group of navigational neurons; depending on which way the animal is pointing its head, different groups of these cells fire, letting the animal know which way it faces (Taube 1998).

Recent works of E.I. Moser and his colleagues have clarified the role of the dorsocaudal medial entorhinal cortex (dMEC) in the creation of the cognitive maps in rodents, and presumably people. It appears that the dMEC contains a directionally oriented, topographically organized neural map of the spatial environment (Hafting et al. 2005). The key unit of this map is the grid cell, which is activated whenever an animal's position coincides with any vertex of a regularly tessellating grid of equilateral triangles spanning the surface of the environment covered by the animal. The map is anchored to external landmarks, but persists in their absence, suggesting that grid cells are part of a generalized map of space. Within a diameter of a few hundred micrometers or less, the complete range of positions and distances appear to be represented.

The striking topographic organization of grid cells in the dMEC is expressed in a number of metric properties, including spacing, orientation (direction) and field size which were almost invariant at individual recording locations. When the environment was expanded, the number of activity nodes increased, while their density remained constant. The stability of the grid vertices across successive trials of the experiment reported in Hafting et al. (2005) suggests that external landmarks exert a significant influence supporting the notion that phase and orientation of the grid are set by external landmarks. The representation of place, distance and direction in the same network of dMEC neurons permits the computation of a continuously updated position vector of the animal's location (Sargolini et al. 2006). Interestingly, the entorhinal cortex is found upstream of the hippocampus, which suggests that place cells could be learned from the activity of grid cells.

Hippocampal place representations may be derived from a metric representation of space in the medial entorhinal cortex (MEC) (Hafting et al. 2005, Sargolini et al. 2006, McNaughton et al. 2006). Strong evidence indicates that these neurons are part of a path integration system. Local ensembles of grid cells have a rigid spatial phase relationship, so that the grid network (Fig. 1.9) provides a universal metric for path integration-based navigation in space. The ensembles of place cells undergo extensive remapping in response to changes in the sensory inputs to the hippocampus when the animal, for example, alternates the enclosures (Fig. 1.9). Dynamics of cells in the dMEC grid were found to be strongly predictive of the type of remapping induced in the hippocampus. Grid fields of co-localized cells in MEC move and rotate in concert during this remapping, thus maintaining a constant spatial phase structure, allowing position to be represented and updated by the same translation mechanism in all environments encountered by the animal (Fyhn et al. 2007). The grid spacing, grid orientation and spatial phase distribution were found to be preserved between the conditions. Anchoring the output of the path integration to external reference points stored in the hippocampus may enable alignment of entorhinal maps, whatever departure point is chosen.

A spatial map based on the Euclidean metric approximating the regularly tessellating grid of equilateral triangles and anchored to external landmarks noticed and remembered because of dominance of visible forms, or because of sociocultural significance, would probably be the best representation of environments in humans. Then, the path integration mechanism over regular grids of cells with different orientations and spacings by which information about distance and direction

Fig. 1.9 The grid cells of the dMEC in rats show higher firing rates when the position of the animal correlates with the vertices of regular triangular tessellations covering the environment. Strong evidence indicates that these neurons are part of a path integration system

of motion of the individual may be extracted is responsible for the structural understanding of environment, the basic aptitude which is necessary to any creative spatial intervention.

These recent results from neurobiology provide a scientific base for the studies in wayfinding behavior investigating the properties and organization of cognitive maps of space in humans intensively developed by city planners and architects (see Golledge 1999 and references therein).

1.2.3 Space-Based Representations of Urban Environments. Least Line Graphs

In his book "The Image of the City," Lynch (1960) presented the results of his study on how people perceive and organize spatial information as they navigate through cities.

In his theoretical description of the city's visual perception grounded on objective criteria, he introduced innovative concepts of place legibility (the ease with which city layouts are recognized) and imageability, which had later been used in the Geographic Information Systems (GIS) development. He suggested that well-designed paths relying on the clarity of direction are significant not only for pursuing the practical tasks such as wayfinding, but also are central to the emotional and physical well-being of the inhabitant population, personally as well as socially. Lynch's study was intended to develop a general method for mapping the city in terms of its most significant or imaginable elements.

Open spaces of the city may be broken down into components; most simply, these might be straight street segments tracing over every longest line of sight, which then can be linked into a network via their intersections and analyzed as a network of movement choices.

A set of theories and techniques called space syntax had been conceived by B. Hillier and colleagues at the University College of London in the late 1970s for the analysis of spatial configurations establishing relations between all open spaces in an urban environment (Hillier 1996). Being developed as a tool to help architects simulate the likely social effects of their designs, the approach became instrumental in predicting human behavior i.e, pedestrian movements in urban environments (Jiang 1998). It has been shown within the space syntax approach that, when seen as configuration, space in cities is not a neutral background for physical entities, but in itself has both structure and agency (Hillier 2008).

Syntactic apprehension of space is by no means new; people have implemented it intuitively for one thousand years in naive geography, in which topological relationships between places were used prior to any precise measurement. The reason for an everyday success of naive geography is quite simple: human thinking and perception of places are not simply metric, but are rather based on the perception of vista spaces as single units and on the understanding of topological relationships between these vista spaces.

The nature of the human perception of places became clear after the introduction of spatial network analysis software. It had been quickly recognized that there was a big gap between what a human user wants to do with a GIS and the spatial concepts originally offered by the GIS programmers (Egenhofer et al. 1995). Long ago the American Automobile Association developed a strip map representing a route by a sequence of concatenated segments, each end anchored by a significant choice point. Today, it is common to use the strip map representations in onboard navigation computer systems.

Open spaces are all interconnected, so that one can travel within them to and from everywhere in the city. It is sometimes difficult to decide what an appropriate spatial element of the complex space involving large numbers of open areas and many interconnected paths should be.

To deal with street connectivity within the standard space syntax approach, all straight lines in the plan that are tangent to pairs of city block vertices and extend until either the line is incident on a block, or on a notional boundary, can be drawn around the urban pattern to represent its limits. Indeed, the resulting dense array of lines contains many subsidiary lines whose set of connection is a subset of those of another line. An elimination algorithm proposed by Turner (2003) allows the smallest set of lines that cover all the space and make all connections from one line to another called the least line map (graph). This map is then considered as a graph in which the lines are nodes and intersections are links subjected to further analysis and tests. In contrast to the object-based graph representations of the city, in general the least line graph is not planar. Providing a fundamental, evidence-based approach to the planning and design of buildings and cities, the space syntax analysis is rather time consuming for large networks.

Moreover, it has been pointed by Ratti (2004) that the traditional axial technique exhibits inconsistencies since the use of straight lines is oversensitive to small deformations of the grid, which leads to noticeably different graphs for systems that should have similar configuration properties. In particular, it has been recently pointed out by Figueiredo et al. (2007) that in the framework of traditional space syntax techniques there is an artificial differentiation between straight and curved or sinuous paths that have the same importance in the system. Long straight paths, represented as a single line, are overvalued compared to curved or sinuous paths as they are broken into a number of straight axial lines.

In the work of Jiang et al. (2004) it has been suggested that the nodes of a morphological graph representing the individual open spaces in the spatial network of an urban environment should have an individual meaning, and that the hierarchy and geometry of the system should be encapsulated in the structure of the graph. While identifying a street over a plurality of routes on a city map, the named-street approach has been used by Jiang et al. (2004) in which two different arcs of the original street network were assigned to the same street identification number (ID) provided they have the same street name. The main problem of the approach is that the meaning of a street name could vary from one district or quarter to another even within the same city. For instance, the streets in Manhattan do not meet, in general, the continuity principle, rather playing the role of local geographical coordinates.

In Figueiredo et al. (2005), two axial lines were aggregated if and only if the angle of continuity between the linear continuation of the first line and the second line was less than or equal to a pre-defined threshold. If more than one continuation was available, the line corresponding to the smaller angle was chosen.

In Porta et al. (2006), an intersection continuity principle (ICN) has been proposed accordingly so that two street segments forming the largest convex angle in a junction on the city map are assigned the highest continuity and, therefore, are coupled together, acquiring the same street ID. The main problem with the ICN principle is that the streets crossing under convex angles would artificially exchange their identifiers, which is not crucial to the study of the probability degree statistics performed by Porta et al. (2006), but makes it difficult to interpret the results if the dynamical modularity of the city is studied (Volchenkov et al. 2007a). It is also important to mention that the number of street IDs identified within the ICN principle usually exceeds the actual number of street names in a city.

In Cardillo et al. (2006), Scellato et al. (2006) and Crucitti et al. (2006), the ICN principle has been implemented in order to investigate the relative probability degree statistics and some centrality measures in the spatial networks of a number of the one square mile representative samples taken from different cities of the world. However, the decision about which square mile would provide an adequate representation of a city is always questionable.

Recently, it has been pointed out by Turner (2007) that all above-mentioned methods of angular segment analysis of city space syntax can marry the traditional axial and road-center line representations through a simple length-weighted normalization procedure that makes values between the two maps comparable.

1.2.4 Time-based Representations of Urban Environments

The time-based representation of urban spatial networks is established on the ideas of traffic engineering and queueing theory invented by A. K. Erlang (see Brockmeyer et al. 1948). It arises naturally when we are interested in how much time a walker or a vehicle would spend travelling through a particular place in the city.

The common attribute of all spaces of motion in the city is that we can spend some time while moving through them. All such spaces found in the city are considered to be physically identical, so that we can regard them as nodes of a queueing graph $G(V,E)$, in which V is the set of all spaces of motion, and E is the set of all their interconnections.

Every space of motion $i \in V$ is considered as a service station of a queueing network (QN) (Breuer et al. 2005) characterized by some time of service, so that the relations between these service stations – the segments of streets, squares, and roundabouts – are also traced through their junctions. Travellers arriving to a place accordingly to some interarrival time distribution are either moving through it immediately or queueing until the space becomes available. Once the place is passed through, the traveller is routed to its next station, which is chosen randomly in accordance with a certain probability distribution among all other open spaces linked to the given one in the urban environment. However, if the destination space has finite capacity, then it may be full and the traveller will be blocked at the current location until the next space becomes available.

The paths along which a traveller may move from service station to service station are considered in queueing theory as being random and determined by the routing probabilities, so that the theory of Markov chains (see Markov 1971) provides the generative statistical models for the analysis of QN (Bolch et al. 2006).

In contrast to the object-based and space-based representations discussed above, the urban spatial network considered within the time-based approach is a QN of N interconnected servers and, therefore, its adequate time-based representation depends essentially upon the overall utilization of space in the city.

How many servers do we need in order to represent the urban spatial network as the queueing network reliably and consistently?

To answer this question in the spirit of queueing theory, we should suggest that the arrival rate of travellers per unit time Λ and the service rate per unit time $\Omega \ll \Lambda$ are uniformly fixed for all open spaces in the city. Then the minimum number of servers needed to represent the urban spatial network can be estimated simply by

$$N \geq \frac{\Lambda}{\Omega}. \tag{1.3}$$

Although, this number may be very large, it should obviously be much smaller than the total number of travellers through the spatial network. Similarly to the spatial graph representations of urban environments studied within the space syntax approach, the queueing graphs are not planar.

It is important to mention that some components of the city spatial network may induce specific symmetric fully connected subgraphs (cliques) into the single nodes with summed time of service.

The computational task of determining whether a graph contains a clique (the clique problem) is known to be a graph-theoretical NP-complete problem,[1] (Karp 1972). A standard algorithm to find a clique in a graph is to start by considering each node to be a clique of size one, and to merge cliques into larger cliques until there are no more possible merges. Two cliques may be merged if each node in the first clique is adjacent to each node in the second clique. Although this algorithm can fail to find the essentially large cliques, it can be improved using the well-known union-find algorithm (Cormen et al. 2001).

In the context of QN, the waiting time probability distribution is a central quantity characterizing the network dynamics.

---

[1] In computational complexity theory a decision problem belongs to the class NP (nondeterministic polynomial time) if it can be decided by a nondeterministic Turing machine, an idealized model for mathematical calculation, in polynomial time. A problem is said to be NP-hard if any NP problem can be deterministically reduced to it in polynomial time. A problem is said to be NP-complete if it is both NP and NP-hard.

It should be noted that various models of human behavior based on the QN principles are widely used in modelling traffic flow patterns or accident frequencies (Haight 1967), and are commercially used in call center staffing (Reynolds 2003), inventory control (Greene 1997), or to estimate the number of congestion-caused blocked calls in mobile communication (Anderson 2003). In all these models it is assumed that human actions are randomly distributed in time and thus well approximated by Poisson processes predicting that the time interval between two consecutive actions by the same individual, called the waiting or interevent time, follows an exponential distribution (Haight 1967).

Recently, it has been pointed out by (Barabasi 2005) that the fact that humans assign their active tasks and future actions different priorities may lead to human activity patterns which display a bursty dynamic with interevent times following a heavy tailed distribution. A relevant process that can be modelled as a priority queueing system in which tasks arrive randomly and require the processing action of the human has been discussed in detail by Vasquez (2005), Vasquez et al. (2006) and Blanchard et al. (2007). In particular, it has been demonstrated that fat tails for the waiting time distributions are induced by the waiting times of very low priority tasks that stay unserved almost forever as the task priority indices are "frozen in time" (i.e., a task priority is assigned once for all to each incoming task). However, when the priority of each incoming task is time-dependent, and "aging priority mechanisms", which ultimately assign high priority to any long waiting tasks, are allowed then fat tails in the waiting time distributions cannot find their origin in the scheduling rule alone (Blanchard et al. 2007). These results may have important implications for understanding the traffic dynamics in urban area networks.

1.2.5 How Did We Map Urban Environments?

Provided all individual spaces of motion are taken as physically identical and, in particular, their service rates per unit time are uniformly equal, then, for all practical purposes, a simple heuristic can be used in order to identify the valuable morphological components of urban environments.

Squares and roundabouts are very important objects for the city morphology. In Lynch (1960) the squares have been treated as the basic elements among those which help people to perceive spatial information about the city. It is remarkable that squares and roundabouts broken into a number of equivalent "service stations" forming a symmetric subgraph always acquire the individual identification number (ID) as the result of shrinking.

The subsequent encoding of the city spatial network into a graph was the same as in the traditional space syntax techniques. Nodes represented the spaces of motion characterized by some travelling times, and edges stayed for their overlaps. In the setting of discrete time stochastic models, instead of specific travelling time for each node $i \in V$ of the graph G, we defined the laziness parameter $\beta_i \in [0, 1]$ which quantified the probability that a random walker leaves the node in one time step, while $1 - \beta_i$ equals the probability the walker stays in i. We assigned an individual street ID code to each continuous part of a street even if all of them share the same street name. Then the spatial graph was constructed by mapping edges of the primary graph encoded by the same street IDs into nodes and intersections among each pair of edges in the primary graph into edges connecting the corresponding nodes of the temporal graph.

It is clear, assuming all spaces of motion in the city are equal, the time-based representation algorithm in general creates the same spatial graph as obtained by the standard street-named approach suggested by Cardillo et al. (2006), Porta et al. (2006), Scellato et al. (2006), and Crucitti et al. (2006); however, the possible discontinuities of streets are also taken into account. Namely, each continuous part of a street acquires an individual street ID code in the temporal graph even if all of them share the same street name. This heuristic approach has been used to analyze urban environments in Volchenkov et al. (2007a) and Volchenkov et al. (2007b). It is also important to mention that the spatial graph of urban environments planned in grids are essentially similar to those of dual information representation of the city map introduced earlier by Rosvall et al. (2005).

The transition from the city plan to its spatial graph representation is a highly nontrivial topological transformation of a planar graph into a non-planar one that encapsulates the hierarchy and structure of the urban area. Below, we present a glossary establishing a correspondence between the typical components of urban environments and certain elements of spatial graphs.

The topological transformation replaces the one-dimensional open segments (streets) by the zero-dimensional nodes (Fig. 1.10(1)). The sprawl-like developments consisting of a number of blind passes branching off a main route are changed to the star subgraphs having a hub and a number of client nodes (Fig. 1.10(2)). Junctions and crossroads are replaced with edges connecting the corresponding nodes of the dual graph (Fig. 1.10(3)). Squares and roundabouts are considered independent topological objects and acquire the individual IDs (Fig. 1.10(4)). Cycles are converted into cycles of the same lengths (Fig. 1.10(5)). A regular grid pattern shown in (Fig. 1.10(6)) is replaced by a complete bipartite graph, where the set of vertices can be divided into two disjoint subsets such that no edge has both end points in the same subset, and every line joining the two subsets is present (Krueger 1989). These disjoint sets of vertices in the bipartite graph can be naturally interpreted as the vertical and horizontal edges, respectively (i.e., streets and avenues).

In the work of Brettel (2006) it has been established, using the space syntax approach, that people's perception of a neighborhood and choice of routes depends upon the order in a street layout that represents similar geometrical elements in repetition. It is the spatial graph transformation which allows separating the effects of order and of structure while analyzing the spatial network on the morphological ground. It converts the repeating geometrical elements expressing the order found in the urban developments into the twin nodes, the pairs of nodes such that any other is adjacent either to them both or to neither of them. Examples of twin nodes can be found in Fig. 1.10(2,4,5,and 6).

Twin nodes correspond to the multiple eigenvalue $\lambda = 1$ of the normalized Laplace operator defined on the spatial graph and to the multiple eigenvalue $\mu = 0$ of the Markov transition operator of random walks (see Chapter 2). Therefore, all similar repeating geometrical elements in the urban spatial network contribute one and the same eigenmodes of the passive transport processes defined on the corresponding spatial graph.

1.3 Structure of City Spatial Graphs

Most real world networks can be considered complex by virtue of features that do not occur in simple networks.

If cities were perfect grids where all lines have equal lengths and the same number of junctions, they would be described by regular graphs exhibiting a high level of

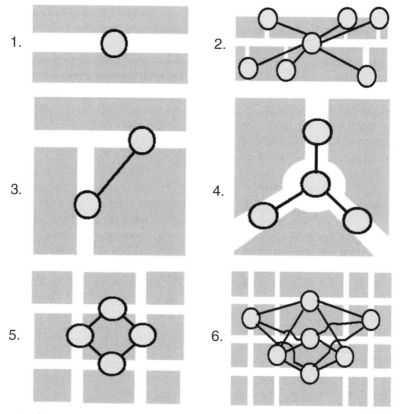

Fig. 1.10 The topological transformation glossary between the typical components of urban environments and the certain elements of temporal graphs

similarity no matter which part of urban texture is examined. It has been suggested in Rosvall et al. (2005) that a pure grid system is easy to navigate since it provides the multiple routes between any pair of locations and, therefore, minimizes the number of necessary navigation instructions. Although, the perfect urban grid minimizes descriptions, its morphology does not differentiate the main spaces, so that movement tends to be dispersed everywhere since, in the ideal grid, all routes are equally probable (Figueiredo et al. 2007).

Alternatively, if cities were purely hierarchical systems (like trees), they would clearly have a main space (a hub, a single route between many pairs of locations) that connects all branches and controls movement between them. This would create a highly segregated, sprawl-like system that would cause tough social consequences (Figueiredo et al. 2007).

However, cities are neither trees nor perfect grids, but a combination of these structures that emerges from the social and constructive processes (Hillier 1996). They maintain enough differentiation to establish a clear hierarchy (Hanson 1989) resulting from the interplay between the public processes, such as trade and exchanges, and the residential process preserving their traditional structures. The emergent urban network usually possesses a very complex structure which is naturally subjected to the complex network theory analysis.

We assume that all continuous spaces between buildings restraining traffic in the urban pattern are regarded as nodes $V = \{1, \ldots, N\}$ of the temporal graph $G(V, E)$, in which any pair of individual spaces, $i \in V$ and $j \in V$, are held to be adjacent, $i \sim j$, when it is possible to move freely from one space to another, without passing through any intervening. The space adjacency relations between all nodes are encoded by edges, $(i, j) \in E$, if and only if $i \sim j$.

1.3.1 Matrix Representation of a Graph

Although graphs are usually shown diagrammatically, they can also be represented as matrices. The major advantage of matrix representation is that the analysis of graph structure can be performed using well-known operations on matrices. For each graph, there is a unique adjacency matrix (up to permuting rows and columns) which is not the adjacency matrix of any other graph.

If we assume that the spatial graph of the city is simple (i.e., it contains neither loops, nor multiple edges), the adjacency matrix is a $\{0, 1\}$-matrix with zeros on its diagonal:

$$A_{ij} = \begin{cases} 1, & i \sim j, \quad i \neq j, \\ 0, & \text{otherwise.} \end{cases} \qquad (1.4)$$

If the graph is undirected, the adjacency matrix is symmetric, $A_{ij} = A_{ji}$. If the graph contains twin nodes, the corresponding rows and columns of $\mathbf{A}$ are identical.

In graph theory, connectivity of a node is defined as the number of edges connected to it in the graph. The degree of a node $i \in V$ is the number of other nodes adjacent to i in G,

$$\deg(i) = \mathrm{card}\{j \in V : i \sim j\}$$
$$= \Sigma_{j \in V} A_{ij}. \tag{1.5}$$

Degree is often interpreted in terms of the immediate risk of node for catching whatever is flowing through the network (such as a virus, or some information). Higher connectivity nodes play the role of hubs being traversed by more paths between various origin/destination pairs than places with less connectivity.

City districts constructed in accordance with different development principles and in different historical epochs may be identified and visualized on a three-dimensional representation of the city spatial graph (Volchenkov et al. 2007a).

In Fig. 1.11, we show the city map of Bielefeld downtown (left), in which part A kept its original structure founded in the 13th–14th Centuries, while part B was subjected to the partial redevelopment in the 19th Century. On the right side of Fig. 1.11, the three-dimensional representation of the spatial graph of the Bielefeld downtown is presented. The (x_i, y_i, z_i)-coordinates of the ith vertex of the spatial graph in three-dimensional space are given by the relevant ith-components of three eigenvectors (u_2, u_3, u_4) of the adjacency matrix $\mathbf{A}$. These eigenvectors correspond to the second, third, and fourth largest (in absolute value) eigenvalues of $\mathbf{A}$. The three-dimensional spatial graph of Bielefeld clearly displays a structural difference between parts A and B in the three-dimensional representation, the relevant subgraphs are located in the orthogonal planes of Fig. 1.11 (right). Sometimes other

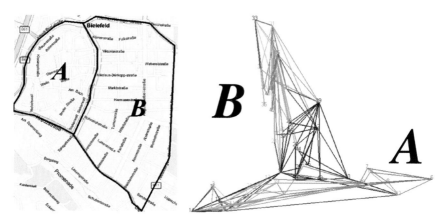

Fig. 1.11 The city map of Bielefeld downtown (*left*) is presented together with its three-dimensional representations of the spatial graph (*right*). The A part keeps its original structure (founded in the 13th–14th Centuries); part B which had been redeveloped in the 19th Century. The (x_i, y_i, z_i)-coordinates of the ith vertex of the spatial graph in three-dimensional space are given by the relevant ith-components of three eigenvectors (u_2, u_3, u_4) of the adjacency matrix $\mathbf{A}$ of the spatial graph

symmetries of spatial graphs can be discovered visually by using other triples of eigenvectors if the number of nodes in the graph is not very large.

In space syntax theory, a number of configuration measures have been introduced in order to describe the quantitative representations of relationships between spatial components in the city. Among them, the control value (CV) of a space evaluates the scale to which a space controls access to its immediate neighbors, taking into account the number of alternative connections that each of these neighbors has (Jiang 1998),

$$
\begin{aligned}
\mathrm{CV}(i) &= \Sigma_{i \sim j} \frac{1}{\deg(j)} \\
&= \Sigma_{j=1}^{N} \left(\mathbf{A}\mathbf{D}^{-1} \right)_{ij}
\end{aligned}
\tag{1.6}
$$

where the diagonal matrix $\mathbf{D} = \mathrm{diag}\left(\deg(1), \deg(2), \ldots, \deg(N)\right)$. The control value of a node is closely related to the concept of random walks estimating the degree of choice the node represents for others directly linked to it (Jiang et al. 2000).

There are simple facts about adjacency matrices that we mention here, while the details can be found in Godsil et al. (2004). If $\mathbf{A}$ is the adjacency matrix of the graph G, then the entry in row i and column j of its nth power $\mathbf{A}^n$ gives the number of paths of length n from vertex i to vertex j. The matrix $\mathbf{1} - \mathbf{A}$ (where $\mathbf{1}$ denotes the $N \times N$ identity matrix) is invertible if and only if there are no directed cycles in the graph G. In this case, the entry in row i and column j of the inverse $(\mathbf{1} - \mathbf{A})^{-1}$ gives the number of paths from vertex i to vertex j (which is always finite if there are no directed cycles in G).

The number of cycles of length l in the graph G can be accounted with the help of the polynomial $\det(\mathbf{1} - z\mathbf{A})$ that has been investigated by Ihara (1966). The number C_l of l-cycles is given by

$$
C_l = \frac{1}{(l-1)!} \frac{d^l}{dz^l} \frac{1}{\det(\mathbf{1} - z\mathbf{A})} \bigg|_{z=0},
\tag{1.7}
$$

in particular, $C_1 = 0$ if the graph has no self-loops.

1.3.2 Shortest Paths in a Graph

In graph theory, the shortest path problem consists of finding the quickest way to get from one location to another on a graph. The number of edges in a shortest path connecting two vertices, $i \in V$ and $j \in V$, in the graph is the distance (also: depth) between them, d_{ij}. It is obvious that $d_{ij} = 1$ if $i \sim j$.

There are a number of other graph properties defined in terms of distance. The diameter of a graph is the greatest distance between any two vertices, $\mathfrak{D}_G = \max_{i,j \in V} d_{ij}$. The radius of a graph is given by $\mathfrak{R}_G = \min_{i \in V} \max_{j \in V} d_{ij}$.

The distance would determine the relative importance of a vertex within the graph. The mean distance (also: mean depth) from vertex i to any other vertex in the graph is

$$\ell_i = \frac{1}{N-1} \sum_{j \in V} d_{ij}.$$ (1.8)

In space syntax theory, the mean depth (1.8) is used to quantify the level of integration/segregation of the given location from the rest (Jiang 1998). A location is more integrated if all the other places can be reached after traversing a small number of intervening places. The more integrated places are likely to be featured on more topologically short routes than others and are, on average, more accessible than others. In particular, well-integrated streets can account for approximately 70 to 75 percent of pedestrian flow rates in some urban environments (Dalton 2007). Alternatively, a place is said to be less integrated if the necessary number of intermediate locations increases. In graph theory, the relation (1.8) expresses the closeness as a centrality measure of the vertex within a graph. The more integrated nodes that tend to have short geodesic distances to others have higher closeness (the "shallow" nodes).

Betweenness is another centrality measure of a vertex within a graph. In space syntax theory, the betweenness of the node is called the global choice, a dynamic global measure of the flow through a space (Hillier et al. 1987). It captures how often, on average, a location may be used in journeys from all places to all others in the city. Locations that occur on many of the shortest paths between others (i.e., provide a strong choice) have higher betweenness than those that do not. Global choice is estimated as the ratio

$$\text{Choice}(i) = \frac{\{\#\text{shortest paths through } i\}}{\{\#\text{all shortest paths}\}}.$$ (1.9)

Betweenness is, in some sense, a measure of the influence a node has over the spread of information through the network. The betweenness centrality index is essential in the analysis of many real world networks and social networks, in particular, but costly to compute. The Dijkstra algorithm and the Floyd-Warshall algorithm (Cormen et al. 2001), may be used in order to calculate $\text{Choice}(i)$.

1.3.3 Degree Statistics of Urban Spatial Networks

In the framework of complex network theory, the focus of study is shifted away from the analysis of properties of individual vertices to consider the statistical properties of graphs which are supposed to be very large (Newman 2003).

The probability degree distribution

$$P(k) = \Pr[i \in G \mid \deg(i) = k]$$ (1.10)

is nothing else but the probability that a node selected at random among all nodes of the graph has exactly k neighbors. The importance of highly connected nodes for complex social and biological networks of all kinds was recognized long ago (Rapoport 1957).

Since the early study of random graphs performed by Erdös et al. (1959), the degree distribution has become a common way of classifying very large graphs into categories, such as random graphs, for which $P(k)$ asymptotically follows a Poisson distribution, or scale-free graphs (Barabasi et al. 1999), for which

$$\Pr[\deg(i) = k] \simeq \frac{1}{k^\gamma}, \quad \gamma > 1. \tag{1.11}$$

Structure and dynamics of many immense social, information, and transport networks are apparently independent of the network size N. The probability degree distribution observed for many of them follows a power law (1.11) with some exponent $\gamma > 1$.

The importance of the implemented street identification principle is worth mentioning for the investigations on the degree statistics of the spatial city networks.

In general, compact city patterns do not provide us with sufficient data to draw conclusions about the universality of degree statistics. The comparative investigations of different street patterns performed in Cardillo et al. (2006) and Porta et al. (2006) reveal scale-free degree distributions for the vertices of spatial graphs if the ICN principle is implemented. However, it has been reported by Jiang et al. (2004) that, under the street name approach, the spatial graphs exhibit small world character, but scale-free degree statistics can hardly be recognized.

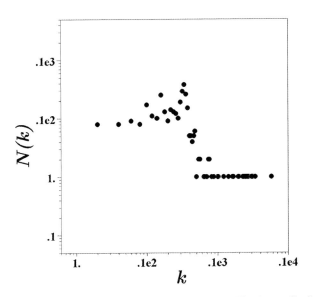

Fig. 1.12 The log-log plot of the empirical probability degree distribution observed for the spatial graph of Manhattan. Data points indicate the numbers of streets in Manhattan, $N(k)$, versus the number k of junctions they have with others

To give an example, we display in Fig. 1.12 the log-log plot of the empirical probability degree distribution observed for the spatial graph of Manhattan which is obviously scale-dependent. It is remarkable that the empirical probability degree distributions observed for the spatial graphs of compact urban patterns are usually broad indicating that the itineraries can cross different numbers of others. Nevertheless, these distributions usually have a clearly recognizable maximum corresponding to the most probable number of junctions an average route has in the city. The distributions usually exhibit a long right tail that decays faster then any power law due to just a few routes that cross many more others than the average.

However, for the relatively large temporal graphs representing the urban environments of megacities probably containing many thousands of different locations, a power law tail can be observed in the probability degree statistics. In Fig. 1.13, we have shown the log-log plot of the number of locations in the spatial graph of Paris (which encodes 5,131 locations and 11,796 of their overlaps enclosed by the Peripheral Boulevard) versus the number of their junctions with others. The spatial network of Paris forms a highly heterogeneous graph.

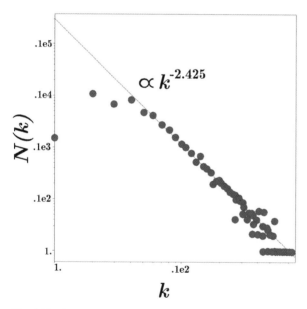

Fig. 1.13 The log-log plot represents the empirical degree statistics for the spatial graph of Paris enclosed by the Peripheral Boulevard (of 5,131 nodes and 11,796 edges). $N(k)$ is the number of locations adjacent to exactly k other locations on the map of Paris. The slope of the regression line equals 2.425

1.3.4 Integration Statistics of Urban Spatial Networks

The measure of integration introduced by Hillier et al. (1984) is one of the principal empirical measures used in the space syntactic characterization of urban morphology and in the pedestrian volume modelling (Raford et al. 2003). Being a typical graph-theoretic centrality measure, it evaluates how many other nodes can be reached from the given one in just a few steps. Most locations in the urban spatial network can be reached immediately from the strongly integrated places.

It is, therefore, reasonable to ask how many locations in the city are relatively strongly integrated and how many others are less integrated and probably belonging to the city fringe.

The degree of a place integration in the spatial graph of the city containing N different locations is quantified by the Relative Asymmetry index (RA) introduced in Hillier et al. (1984) as the mean distance (1.8) of the node normalized by that in the complete graph $\mathbb{K}_N$ containing N nodes,

$$\mathrm{RA}(i) = 2\frac{\ell_i - 1}{N - 2}. \tag{1.12}$$

It is important to note that independently of N, the value of $\mathrm{RA}(i) \in [0,1]$.

The probability integration distribution,

$$P(\mathrm{x}) = \Pr\left[i \in V | \mathrm{RA}(i) = x\right], \quad x \in [0,1], \tag{1.13}$$

gives the probability for a randomly selected place to have integration $\mathrm{RA}(i) = x$.

The probability integration distribution calculated for two German medieval organic cities unveils their structural difference (see Fig. 1.14). The probability integration distributions for organic cities are typically bell-shaped (see the probability integration distribution for the city of Rothenburg indicated by the dashed line in Fig. 1.14), however may exhibit a relatively long right tail due to the strongly integrated itineraries forming the city core. The temporal routes which usually branch out from the major itineraries and penetrate deeply inside the quarters contribute to the city bulk forming the maximum of the distribution. Eventually, a left tail of the probability integration distributions signals about the loosely integrated locations belonging either to the city fringe, or to the relatively segregated neighborhoods.

The right inset in Fig. 1.14 displays the three-dimensional image of the spatial graph for Rothenburg. The (x_i, y_i, z_i)-coordinates of the i-th vertex of the spatial graph in three-dimensional space are given by the relevant i-th components of three consequent eigenvectors u_2, u_3, and u_4 of the adjacency matrix $\mathbf{A}_R$ for the spatial graph of Rothenburg. The spatial graph is supposed to be undirected and, therefore, its adjacency matrix is symmetric, having real eigenvectors and an orthonormal system of eigenvectors. The eigenvectors u_2, u_3, u_4 correspond to the second, third, and fourth largest eigenvalues (in absolute value) of the adjacency matrix. In contrast to the diamond looking three-dimensional image of the spatial graph for Rothenburg, the graph for the downtown of Bielefeld shown in the left inset of Fig. 1.14 reveals the structural heterogeneity of the urban pattern.

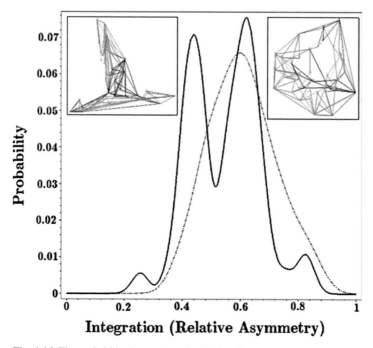

Fig. 1.14 The probability integration distribution for the downtown of Bielefeld (*the solid line*) and for the Rothenburg (the dashed line). The left inset shows the spatial graph of the downtown of Bielefeld, another inset represents the spatial graph of Rothenburg

Two parts of Bielefeld's downtown had been developed accordingly and different principles in different historical epochs can be easily visualized on the graph given in the left inset of Fig. 1.14. The ancient part of Bielefeld is located on the bottom of the picture, while the redeveloped quarters mostly planned on a regular grid are lifted up and placed on the orthogonal plane.

The striking structural dissimilarity of two parts in the urban pattern of Bielefeld can also be traced by the probability integration distribution profile designated in Fig. 1.14 by the solid line. Two maximums of the Relative Asymmetry distribution inform us about two different most probable values of integration which are always higher for the streets belonging to one and the same city district, but are lower with respect to the streets from the alternative city components.

It is also remarkable that the partial redevelopments performed in the downtown of Bielefeld during the previous centuries had enhanced the effect of ghettoization of the relatively isolated quarters in the ancient part of the city (they became even more isolated for the travellers starting from the modernized urban component - see the temporal maximum on the left side of the probability integration distribution), but, at the same time, reinforced the city core by giving it the crucial role in sustaining the entire city's connectedness. The central city itinerary (Niederwall) being a boundary between two parts of the city conjugates them both (see the temporal peak on the right side).

1.3.5 Scaling and Universality: Between Zipf and Matthew. Morphological Definition of a City

In urban studies, scaling and universality are very common. The evolution of social and economic life in cities increases with the population size: wages, income, gross domestic product, bank deposits, as well as the rate of innovations, measured by the number of new patents and employment in creative sectors scale superlinearly, over different years and nations, with statistically consistent exponents (Florida 2004, Bettencourt et al. 2007). The comparative investigations of different urban patterns in the framework of space syntax approach performed by Hillier (2002), and later by Carvalho et al. (2004), show that the space-objected representations (the least line maps) of cities also have scale-free properties in terms of the line length distributions consisting of a very small number of long lines and a very large number of short lines.

The famous rank-size distribution of city sizes all over the world is known as Zipf's Law (Zipf 1949) (see Sec. 1.1.4). Here we report on the similar rank distributions for the values of the space syntax measures quantifying the centrality of nodes in the spatial graphs calculated for the compact urban patterns (see Figs. 1.15 and 1.16).

We have ranked all open spaces in the spatial graphs of the investigated five compact urban patterns in accordance with their centrality indices. The nodes with the worse centrality have been assigned to the first (lowest) rank, while those of the best centrality have the highest rank. The diagrams for Rank-Centrality distributions have been calculated for all spatial graphs for the urban patterns mentioned in Table 1.1. No matter how the centrality level was estimated, either by the integration values (estimated by the Relative Asymmetry (1.12)) (see Fig. 1.15), or by the global choice values (1.9) (see Fig. 1.16), the data from all centrality indicators demonstrate a surprising universality being fitted perfectly with the linear pattern, when the slope of the line equals 1.

The matching of power law behaviors (the same scaling exponent) can have a deeper origin in the background dynamical process responsible for such a power-law relation. Being diverse in their sizes, forms, economical and political history, cities nevertheless display the identical scaling behavior and probably share similar fundamental mechanisms of open space creation. Formally, such common dynamics can be referred to as universality, so that those cities with the same critical exponents of rank-integration statistics are said to belong to the same universality class.

The observed universality in distributions of centrality values can be used in order to establish the morphological definition of a city. Namely, a human settlement, in which the centrality values distribution satisfies the scaling patterns shown in Figs. 1.15 and 1.16 can be considered as being a city.

The ubiquity of power-law relations in complex systems are often thought to be signatures of hierarchy and robustness. The area distribution of satellite cities around large urban centers has been reported to obey a power-law with exponent $\simeq 2$ (Makse et al. 1995). The fractal dimension of urban aggregates as a global

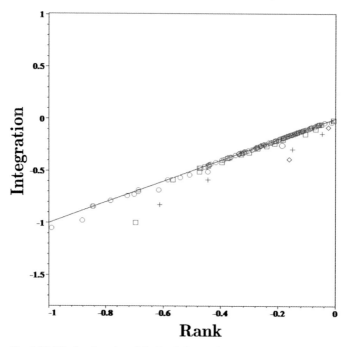

Fig. 1.15 The log-log plot of the Rank-Integration distribution calculated the nodes of five temporal graphs constructed for the street grid of Manhattan, Rothenburg, Bielefeld, the canal networks in Venice and Amsterdam. The linear pattern is given by a straight line

measure of areal coverage have been studied extensively for many cities around the world during the last decades (see Batty et al. 1994 and Tannier et al. 2005) for a review).

In this subsection, we also discuss a scaling property of control values distributions (see Fig. 1.17) calculated for the spatial graphs of compact urban patterns. The $CV(i)$-parameter (1.6) quantifies the degree of choice the node $i \in V$ represents for other nodes directly connected to it.

Provided a random walk, in which a walker moves in one step to another node randomly chosen among all its nearest neighbors is defined on the graph $G(V,E)$, the parameter $CV(i)$ acquires a probabilistic interpretation. Namely, it specifies the expected number of walkers which is found in $i \in V$ after one step if the random walk starts from a uniform configuration, in which all nodes in the graph have been uniformly populated by precisely one walker. Then, the graph G can be characterized by the control value probability distribution

$$P(m) = \Pr[i \in G | CV(i) = m] \tag{1.14}$$

for which the control value of a randomly chosen node equals $m > 0$.

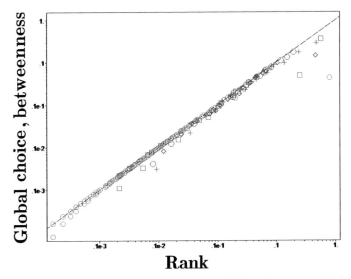

Fig. 1.16 The log–log plot of the Rank-Global Choice (Betweenness) distribution calculated the nodes of five spatial graphs constructed for the street grid of Manhattan, Rothenburg, Bielefeld, the canal networks in Venice and Amsterdam. The linear pattern is given by a straight line

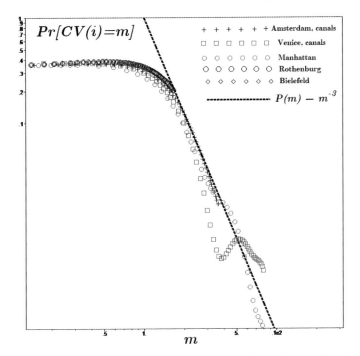

Fig. 1.17 The log–log plot of the probability distribution that a node randomly selected among all nodes of the spatial graph G will be populated with precisely m random walkers in one step starting from the uniform distribution (one random walker at each node). The dashed line indicates the cubic hyperbola decay, $P(m) = m^{-3}$

The log–log plot of (1.14) is shown in Fig. 1.17. It is important to mention that the profile of the probability distribution exhibits a power-law decay for large values $m \gg 1$ well fitted by the cubic hyperbola, $P(m) \simeq m^{-3}$, universally for all five compact urban patterns we have studied.

The universality of statistical behavior of the control values for the nodes representing a relatively strong choice for their nearest neighbors,

$$\Pr[i \in G | \mathrm{CV}(i) = m] \simeq \frac{1}{m^3}, \qquad (1.15)$$

uncovers the universality of the creation mechanism responsible for the appearance of the "strong choice" nodes which act over all cities independently of their backgrounds. It is a common suggestion in space syntax theory that open spaces of strong choice are responsible for the public space processes driven largely by universal social activities like trades and exchange which are common across different cultures and historical epochs and give cities a similar global structure of the deformed wheel (Hillier 2005).

It has been shown long ago by Simon (1955) that power-law distributions always arise when "the rich get richer" principle works i.e., when a quantity increases with its amount already present. In sociology this principle is also known as the Matthew effect (Merton 1968) (this reference appears in (Newman 2003)) following the well-known biblical edict.

1.3.6 Cameo Principle of Scale-Free Urban Developments

It is interesting to consider possible city development algorithms that could lead to the emergence of scaling invariant degree structures in urban environments.

Among the classical models in which the degree distribution of the arising graph satisfies a power-law is the graph generating algorithms based on the preferential attachment approach (Simon 1955). The preferential attachment model of city growth has been discussed by Andersson et al. (2005) where the population of a city proportionates to the number of transport connections with other cities. The nodes of high degrees tend to quickly accumulate even more links representing a strong preference choice for the emerging nodes, while nodes with only a few links are unlikely to be chosen as the destination for a new link. The preferential attachment forms a positive feedback loop in which an initial random degree variation is magnified with time (Albert et al. 2002). It is fascinating that the expected degree distribution in the graph generated in accordance to the algorithm proposed in Barabasi et al. (1999) asymptotically approaches the cubic hyperbola,

$$\Pr[i \in G | \deg(i) = k] \simeq \frac{1}{k^3}. \qquad (1.16)$$

Nevertheless, it is clear that the mechanisms governing the city creation and development certainly do not follow such a simple preferential attachment principle. Indeed, when new streets (public or private) are created as a result of site subdivision or site redevelopment, they can hardly be planned in a way following the preferential attachment principle. The challenge of city modelling calls for the more realistic heuristic principles that could catch the main features of city creation and development.

A prominent model describing urban development should take into account the structure of embedding physical space: the size and shape of the landscape, and the local land use patterns. A suitable algorithm describing development of complex networks which takes into account the affinity toward a certain place has been recently proposed by Blanchard et al. (2004). It is called the Cameo-principle named for the attractiveness, rareness and beauty of the small medallion with a profiled head in relief called Cameo. It is exactly their rareness and beauty which gives them their high value.

In the Economics of Location theory introduced by Loesch (1954) and developed by Henderson (1974) a city, or even more certainly, a particular district may specialize in the production of a good that can be connected with natural resources, education, policy, or just low salary expenditures. City districts compete among themselves in a city market not necessarily connected with the quantity of their inhabitants. The demand for these products comes into the city district from everywhere and can be considered exogenous. In the Cameo model of Blanchard et al. (2004), the local attractiveness of a site determining the creation of new spaces of motion is specified by a real positive parameter $\omega > 0$. Indeed, it is almost impossible to exactly estimate the actual value $\omega(i)$ for any site $i \in G$ in the urban pattern, since such an estimation can be referred to both the economic and cultural factors that may vary over the different historical epochs and over the certain groups of population.

Therefore, in the framework of a probabilistic approach, it seems natural to consider the value ω as a real positive independent random variable distributed over the vertex set of the graph representation of the site uniformly in accordance to a smooth monotone decreasing probability density function $f(\omega)$.

There are just a few distinguished sites which are much more attractive than an average one in the city, so that the density function f has a right tail for large $\omega \gg \bar{\omega}$ such that $f(\omega) \ll f(\bar{\omega})$.

Each newly created space of motion i (represented by a node in the spatial city graph $G(N)$, containing N nodes) may be arranged in such a way to connect to the already existing space $j \in G(N)$, depending only on its attractiveness $\omega(j)$ and is of the form

$$\Pr[i \sim j \mid \omega(j)] \simeq \frac{1}{N \cdot f^\alpha(\omega(j))} \tag{1.17}$$

with some $\alpha \in (0,1)$. The assumption (1.17) implies that the probability to create the new space adjacent to a space j scales with the rarity of sites characterized with the same attractivity ω as j.

The striking observation under the above assumptions is the emergence of a scale-free degree distribution independent of the choice of distribution $f(\omega)$. Furthermore, the exponent in the asymptotic degree distribution becomes independent of the distribution $f(\omega)$ provided its tail, $f(\omega) \ll f(\bar{\omega})$, decays faster then any power law.

In the model of growing networks proposed in Blanchard et al. (2004), the initial graph G_0 has N_0 vertices, and a new vertex of attractiveness ω taken independently uniformly distributed in accordance to the given density $f(\omega)$ is added to the already existing network at each time click $t \in \mathbb{Z}_+$. Being associated to the graph, the vertex establishes $k_0 > 0$ connections with other vertices already present. All edges are formed accordingly to the Cameo principle (1.17). The main result of Blanchard et al. (2004) concerns the probability distribution that a randomly chosen vertex i which had joined the Cameo graph G at time $\tau > 0$ with attractiveness $\omega(i)$ amasses precisely k links from other vertices which emerge by time $t > \tau$. It is important to note that in the Cameo model the order in which the edges are created plays a role for the fine structure of the graphs. The resulting degree distribution for $t - \tau > k/k_0$ is irrelevant to the concrete form of $f(\omega)$ and reads as follows,

$$\Pr \left[\sum_{j:\ \tau(j) > \tau} 1_{i \sim j} = k \right] \simeq \frac{k_0^{1/\alpha}}{k^{1+1/\alpha+o(1)}} \ln^{1/\alpha} \left(\frac{t}{\tau} \right). \tag{1.18}$$

In order to obtain the asymptotic probability degree distribution for an arbitrary node as $t \to \infty$, it is necessary to sum (1.18) over all $\tau < t$ that gives

$$P(k) \simeq \frac{1}{t} \sum_{0 < \tau < t} \frac{k_0^{1/\alpha}}{k^{1+1/\alpha+o(1)}} \ln^{1/\alpha} \left(\frac{t}{\tau} \right) = \frac{1}{k^{1+1/\alpha+o(1)}}. \tag{1.19}$$

The emergence of the power-law (1.19) demonstrates that graphs with a scale-free degree distribution may appear naturally as the result of a simple edge formation rule based on choices where the probability to chose a vertex with a random affinity parameter ω is proportional to the frequency of its appearance. If the affinity parameter ω is itself power-law as distributed one could also use a direct proportionality to the value ω to still get a scale-free graph.

1.3.7 Trade-Off Models of Urban Sprawl Creation

The Vermont Forum on Sprawl defines urban sprawl as "dispersed development outside of compact urban and village centers along highways and in rural countryside" (the quote appears in Dalton (2005)). The term urban sprawl generally has negative connotations due to the health and environmental issues that sprawl creates (Norman et al. 2006). Residents of sprawling neighborhoods tend to emit more pollution per person and suffer more traffic fatalities. Sprawling urban areas increase car transportation the source of the carbon emissions warming the Earth's atmosphere. Urban sprawl is a growing problem in many countries and in the so-called "smart growth" policies , which became popular in the United States during the 1990s,

trumping such mainstays as education and crime in many polls and contributing to
the election of anti-sprawl politicians (Shaw et al. 2000).

A variety of factors driving the sprawl process are rooted in the desire to realize
new lifestyles in suburban environments, outside the main urban area. Sprawl has
accelerated in response to improved transportation links and enhanced personal mobility (EEA 2006). Evidence suggests that where unplanned, decentralized development dominates, sprawl will occur in a mechanistic way. Otherwise, more compact
forms of urban development appear if growth around the periphery of the city is
coordinated by strong urban policy.

There are currently many indicators of sprawl, but the majority of them are economic or land use related (for instance, low population density is an indicator of
sprawl) rather than intrinsically spatial or morphological.

One of the early studies devoted to the significant reduction of the number of
dead-ends in the urban network, caused by the merging of the cart path and road
networks, suggested that the amount of "ringiness" (Hillier et al. 1984) in the system might serve as a good morphological indicator of sprawl. In Dalton (2005), it
has been suggested that the differences between suburban and urban development
and even sprawl are clearly discernable by the proportion and distribution of cycle
lengths in the spatial graph of the network. Typical suburban developments usually
contain a high ratio of cul-de-sacs, the dead-end streets with only one inlet, together
with just a few entrances accessed from central roads.

However, while observing the highly interlaced patterns of modern housing
subdivisions sprawled out over rural lands at the fringe of many urban areas
in the United States and Canada (see Fig. 1.18), one can see that such subdivisions may contain no cycles at all, being an arborescence offering only a
few places to enter and exit sprawl, causing traffic to use high volume collector
roads.

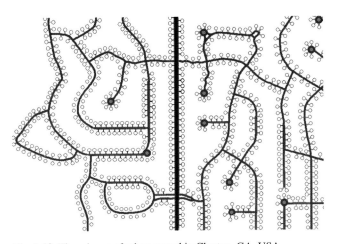

Fig. 1.18 The scheme of urban sprawl in Clayton, GA, USA

Even a brief glance at the satellite image shown in Fig. 1.18 gives the impression that the use of fractal measures can be helpful in the study of urban sprawl. The fractal properties of city sprawl structure is discussed in Batty et al. (1994) and in many other works. Probably, the most fascinatingly reparative morphological element of urban sprawl is individual access from private households to the central path. Designating the individual parking places as the client nodes and the high volume road as a hub in the spatial graph, we obtain a star graph as the typical syntactic motive pertinent to urban sprawl. A star graph consists of a central node (hub) characterized by the uttermost connectivity and a number of terminal vertices linked to the hub.

The temporal graph corresponding to the segment of suburban sprawl shown in Fig. 1.19 forms a star graph, in which 32 individual spaces representing the private parking places are connected to a hub, the only sinuous central road.

Star graphs have been observed in many technical and informational networks. In particular, the algorithms generating the tree-like graphs combined from a number of star subgraphs have been extensively studied and modelled will regard to Internet topology (see Faloutsos et al. (1999), Castells (2001), Fabrikant et al. (2002), and many others).

Extensive experiments suggest that the hierarchical trees containing the numerous star subgraphs could arise as a result of a trade-off process minimizing the weighted sums of two or more objectives.

In a simple model of Internet growth (Fabrikant et al. 2002), a tree is built as nodes arrive uniformly at random in the unit square (the shape is, as usual, inconse-

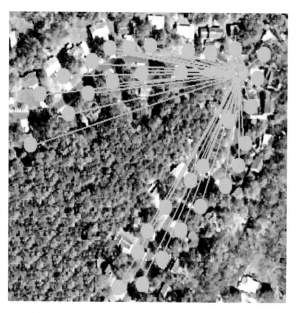

Fig. 1.19 A segment of Atlanta's suburban sprawl (USA). The spatial morphology of sprawl is represented by the star graph

quential). When the i-th node arrives, it connects itself to one of the previous nodes j that minimizes the weighted sum of the two objectives:

$$\text{cost}(i, j) = \alpha \cdot \eth_{ij} + c_j, \quad \alpha \geq 0, \tag{1.20}$$

where $\eth_{ij}$ is the geometrical (Euclidean) distance between two nodes i and j, and c_j is the centrality value of j.

It is newsworthy that the behavior of the trade-off model (1.20) depends crucially on the value of the tuning parameter α which can be naturally interpreted as the "last mile" cost reckoning the construction and maintenance expenditures.

If α is taken less than a particular constant, depending upon the landscape shapes and certain economical conditions, then Euclidean distance is of no importance, and the network produced by the trade-off algorithm is easily seen to form a star.

It seems natural to apply the simple trade-off models to spatial graphs in order to predict the appearance of urban sprawl with the local land use scheme. The decisive factor for the emergence of star graphs is the predominance of the centrality (integration) objective, while the physical (Euclidean) distance between graph vertices is of no importance. It is well-known that the humbleness of physical distances is among main factors shaping the sprawl land use patterns.

The other way around, if we suggest that the last mile costs α grow up with the network size N at least as fast as $\sim \sqrt{N}$, then the Euclidean distance between nodes becomes an important factor shaping the form of the network. A graph that arises in the trade-off process in such a case constitutes a dynamic version of the Euclidean minimum spanning tree, in which high degrees would occur, but with exponentially vanishing probability. By spanning tree we mean a tree containing all the nodes and just enough edges, so that the graph is connected.

Eventually, if α exceeds a certain constant, but grows slower than $\sqrt{N}$, then, as $N \gg 1$, with high probability, the degrees obey a power-law, and the resulting spatial graph forms a fractal.

We suggest that the emergence of a complicated highly inhomogeneous structure which we observe in urban developments can generally involve trade-offs, that is to say the optimization problems between the multiple, complicated and probably conflicting objectives. In order to support this proposition, we have simulated a trade-off model minimizing two different objectives simultaneously. The trading between the geometrical distance and centrality of nodes has been accounted by the objective (1.20). We have also used another cost function,

$$\widetilde{\text{cost}}(i, j) = \omega \cdot \eth_{ij} + d_{ij}, \quad \omega \geq 0, \tag{1.21}$$

in which $\eth_{ij}$ is the Euclidean distance and d_{ij} is the graph theoretical distance between nodes i and j (the number of hopes), as the second objective. In general, the structure of the graph that would arise in such a complex trade-off process crucially depends upon the magnitudes of ω and α, their relative values, and the way they depend on the network size N. In general, it may also depend upon the local curvature of the simulation domain that describes a landscape shape and its distribution.

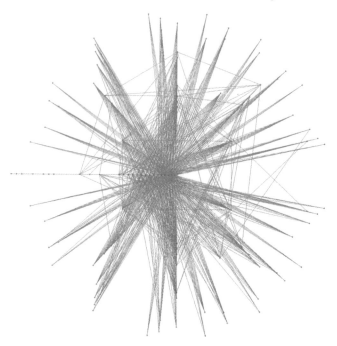

Fig. 1.20 The graph of 500 nodes arisen in the course of the trade-off process minimizing two objective functions simultaneously

In the simplest case of plane domains, for relatively small values of tuning parameters, $\alpha \simeq \omega \simeq 0$, the geometrical distances between nodes are of no importance, and the structure of the resulting graph (shown in Fig. 1.20) is shaped by the simultaneous optimization between the graph distance and centrality objectives. The resulting graph sketched in Fig. 1.20 contains a valuable fraction of twin nodes (while $\alpha = \omega = 10^{-3}$, 84% of graphs nodes are twins) forming almost a complete bipartite morphological graph.

1.4 Comparative Study of Cities as Complex Networks

The traditional comparative studies of cities adopt classifications, such as cultural or geographical criteria, and then apply analytical tools to characterize the existing groups of urban patterns in morphological terms (Major 1997, Karimi 1997, Hillier 2002). However, in a comparative classification of urban textures performed within the space syntax approach, the pre-defined categories have been avoided, and groups have been interpreted as a result of the analysis (Meridos et al. 2005, Figueiredo et al. 2005). A group of methods for automatic classification or grouping, broadly termed "hierarchical clustering" have been discussed in Figueiredo et al. (2005). The general idea behind the hierarchical clustering is that elements of any set have similarities and differences that can be mapped as distances in a multi-dimensional space in which each characteristic (variable) represents an axis. Then, clusters are created by grouping isolated elements or subgroups or, alternatively, splitting the set into smaller groups, according to the distance between them.

In the present section, we discuss the automatic classification methods based on the structural statistics of the nearest and most distant neighbors in the spatial city network.

1.4.1 Urban Structure Matrix

We analyze the spatial graph representing a piece of urban texture from a node (a street or a square) as a starting point. The origin node is immediately adjacent to a number $n_{d=1}$ of other nodes at a distance $d = 1$ from the origin. In their turn, these $n_{d=1}$ nodes are immediately adjacent to $n_{d=2}$ others being located a distance $d = 2$ from the origin node, etc. As a consequence, all locations in the city acquire labels according to the numbers of steps they are apart from the starting place.

In the work of Hillier et al. (1984), such a spatial layout of city locations seen from one of its points has been called the justified graph and widely used in the space syntax analysis of urban environments. The origin location is placed at the bottom of the justified graph as the root space, and all locations that are one step away from the root are put on the first level above, all spaces two steps away on the second level, etc. (Klarqvist 1993) (see Fig. 1.21). It is remarkable that a tree-like justified graph has most of the nodes many steps (levels) away from the root. In such a system the mean distance (also: the mean depth) is large and described as *deep*. A bush-like justified graph has most of the nodes near the root, and the system is described as *shallow*.

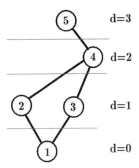

Fig. 1.21 A small justified graph with the root node #1

Any justified graph can be described by the specification vector

$$\left(n_1, n_2, \ldots n_{\mathfrak{D}(G)}\right)_i, \tag{1.22}$$

in which n_d, $d = 1, \ldots \mathfrak{D}(G)$, is the number of nodes being located at a distance d away from the root $i \in V$, and $\mathfrak{D}(G)$ is the diameter of G, the maximal distance observed in G. For example, the specification vector for the justified graph given in Fig. 1.21 is $(2, 1, 1)_1$ where we put the index $(\ldots)_1$ indicating that the graph has been drawn with respect to the root node 1.

We generalize the notion of connectivity by taking into account that all faraway neighbors in a given node have a distance $1 \le d \le \mathfrak{D}(G)$. In particular, the first component of (1.22) is nothing but the degree of the node i. Below, we introduce an automatic method classifying urban textures on the basis of statistics over all specification vectors that can be constructed for the given spatial graph.

Likewise, for all techniques grounded on the analysis of the probability degree distributions popular in complex network theory, the method is insensitive to the structural symmetries the spatial graph can have. This can be promptly demonstrated by the graph shown in Fig. 1.21. It is clear that other four justified graphs which can be constructed in addition to one presented in Fig. 1.21 have the following specification vectors: $(2, 2, 0)_2$, $(2, 2, 0)_3$, $(3, 1, 0)_4$, and $(1, 2, 1)_5$. The first two vectors among them (rooted at the twin nodes 2 and 3) are identical as well as the justification they encode (see Fig. 1.22). Despite its strong influence on various properties of complex networks, the degree statistic is but one of many important characteristics of a graph. Complex networks may possess similar degree distributions yet differ widely in other properties. In order to get more detailed information on the structure of the spatial graph than we could obtain from the probability degree distribution alone, we introduce a rectangular, $\mathfrak{D}(G) \times N$ integer structure matrix $\mathbf{J}$, whose entries are the numbers of nodes that have precisely n neighbors at a distance d,

$$J_{nd} = \{\# \text{ nodes having } n \text{ nodes } at\ a \text{ distance } d\}, \tag{1.23}$$

or, equivalently, the numbers of specification vectors where $n_d = d$. By definition, the matrix $\mathbf{J}$ is independent of a particular layout of street IDs being inherent to the spatial graph of the city. We call $\mathbf{J}$ the urban structure matrix, although it can be constructed for any connected graph independently of its meaning. Such a structure matrix has been proposed by Bagrow et al. (2008) in order to portray any large complex networks.

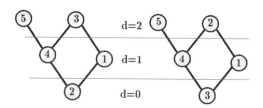

Fig. 1.22 The justified graphs drawn with respect to twin nodes at the root are identical

Fig. 1.23 The matrix plot of the structure matrix for the street network in the downtown of Bielefeld (Westphalia): J_{nd} are the number of streets neighboring precisely to n other streets at a distance d

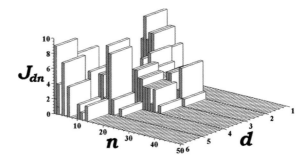

The probability degree distribution is encoded in the first row of the urban structure matrix **J**,

$$J_{1,k} = NP(k), \quad P(k) = \Pr\left[i \in G \,|\, \deg(i) = k\right], \tag{1.24}$$

and other rows of **J** describe the probability distributions for the numbers of faraway neighbors in a node at a distance d,

$$J_{dk} = N \cdot P_d(k), \quad P_d(k) = \Pr\left[i \in G \,|\, (n_d)_i = k\right]. \tag{1.25}$$

In Fig. 1.23, we have presented the urban structure matrix for the spatial graph of the Bielefeld downtown. It indicates a strong scale dependence of probability distributions of faraway neighbors in the spatial graph.

It is remarkable that while the probability degree distribution $P(k)$ of immediate neighbors $(d = 1)$ is a monotonously decreasing function of their numbers, the probability distributions of faraway neighbors $(d > 1)$ shown in Fig. 1.23 are by no means monotonous, so that, for example, it is as much as twice more likely that a street in the Bielefeld center is linked to 23 others in two navigational steps than to 22.

1.4.2 Cumulative Urban Structure Matrix

An alternative way to present the degree statistics is to compute the cumulative probability degree distribution,

$$\mathfrak{P}(k) = \sum_{k'=k}^{N} P(k') \tag{1.26}$$

where $P(k)$ is the probability degree distribution defined by (1.10). $\mathfrak{P}(k)$ is the probability that the degree of a node is greater than or equal to k. The use of $\mathfrak{P}(k)$ instead of $P(k)$ has an advantage, since it reduces the noise in the distribution tail (Newman 2003).

Similarly to (1.26), the statistics of faraway neighbors can also be presented in the form of cumulative distribution functions $\mathfrak{P}_d(n)$ quantifying the probability that the number of neighbors a node has at a distance d is greater than or equal to n. The cumulative distributions $\mathfrak{P}_d(n)$ are the rows of the cumulative urban structure matrix,

$$\mathfrak{I}_{dn} = \frac{\sum_{k \leq n}^{N} J_{dk}}{\sum_{k=1}^{N} J_{dk}}. \tag{1.27}$$

Being a monotonous function of n, the cumulative distribution $\mathfrak{P}_d(n)$ reduces the noise in the distribution tails, however the adjacent points on their plots are not statistically independent.

Usual networks like grids constitute graphs in which most nodes are linked only to their nearest neighbors and, therefore, the total number of faraway neighbors a node has in the graph grows monotonously up as the exploration distance away from the node increases – the far the we reach into the grid, the more neighbors we can achieve.

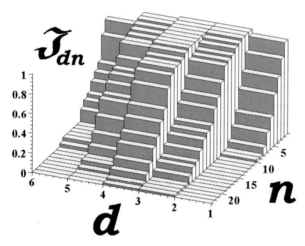

Fig. 1.24 The cumulative urban structure matrix shows the probability distributions of the number n of nearest and faraway neighbors at a distance d in the spatial graph of Rothenburg ob der Tauber (Bavaria). The cumulative probability degree distribution is shown in the first row ($d = 1$). The cumulative distributions of faraway neighbors are encoded by the second and forthcoming rows, $d > 1$

The cumulative urban structure matrix for the spatial graph of the regional railway hub Neubeckum (Westphalia) is shown in Fig. 1.25. The urban area of Neubeckum is bounded by and sprawled along the busiest and most intensively used railway in Germany. It includes the industrial land uses planned for construction and railway maintenance services as well as the residential lands on the vacant lands adjacent to the railway maintenance buildings.

The cumulative urban structure matrices of the spatial graphs of Rothenburg ob der Tauber (Fig. 1.24) and of Venetian canals (Fig. 1.26) show that most of the locations in these urban patterns can be found just three steps away from any other location. In other words, the urban patterns are surprisingly compact containing probably just a few relatively isolated locations a step deeper than others.

The structures of cumulative probability distributions shown in Figs. 1.24, 1.25 and 1.26 show that the correspondent spatial graphs have high representation of subgraphs that miss just a few edges of being cliques.

In complex network theory, a similar phenomenon is referred to as a small world. Small-world networks are characterized by a high clustering coefficient having connections between almost any two nodes within them. Short path lengths are commonly associated with small-world networks (Buchanan 2003).

The shortening of path lengths phenomenon signaled by the cumulative probability distributions shown in Figs. 1.24, 1.25 and 1.26 would originate from the natural mechanism of city developments discussed extensively in space syntax theory. It has been suggested by Hillier (1996) that the spatial structure of organic cities is shaped by the public processes (such as trading and exchange) ordered in such a way as to maximize the presence of people in the central areas. In such a context, the compact structure of the urban patterns unveiled by their cumulative structure matrices uncovers their historical functional pertinence. The tendency to shorten distances in the urban spatial networks induced by public processes is complemented by the residential process which shapes relations between inhabitants and strangers preserving the original residential culture against unsanctioned invasion of privacy. While most streets and canals characterized with an excellent accessibility promote commercial activities and intensify cultural exchange, certain districts in these cities reveal the alternative tendency to segregate the residential areas from the rest of the urban fabric.

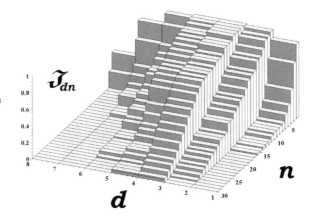

Fig. 1.25 The cumulative urban structure matrix shows the probability distributions of the number n of nearest and faraway neighbors at a distance d in the spatial graph of Neubeckum (Westphalia). The cumulative probability degree distribution is shown in the first row ($d = 1$). The cumulative distributions of faraway neighbors are encoded by the second and forthcoming rows, $d > 1$

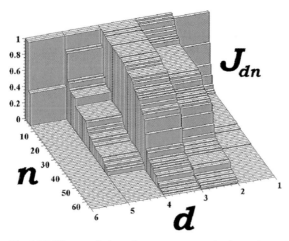

Fig. 1.26 The cumulative urban structure matrix shows the probability distributions of the number n of nearest and faraway neighbors at a distance d in the spatial graph of Venetian canals

1.4.3 Structural Distance Between Cities

We introduce a simple metric comparing two urban patterns by means of their cumulative structure matrices, $\Im$ and $\Im'$. The dissimilarity between these distributions can be measured by Wasserstein's distance widely used in probabilistic contexts (see Rachev 1991). When applied to sample distributions the Wasserstein distance leads to the L_2-based tests of goodness of fit. Given two probability measures, P_1 and P_2, with finite second order moments, the Wasserstein distance between them is defined as the lowest L_2-distance between random variables X_1 and X_2 characterized by the probabilities P_1 and P_2 respectively defined in any probability space, with the following distribution laws:

$$W(P_1, P_2) = \inf \left\{ \sqrt{\mathbb{E}(X_1 - X_2)^2}, \; X_1 \sim P_1, X_2 \sim P_2 \right\}, \tag{1.28}$$

in which $\mathbb{E}(X)$ is the mean of the random variable X. Indeed, the Wasserstein distance is not the only possible measure of dissimilarity between two distributions which can be implemented in the comparative analysis of graph structures. For example, in Bagrow et al. (2008) the Kolmogorov-Smirnov test (Rachev 1991) has been used instead of the Wasserstein distance.

Motivated by the Wasserstein distance, we measure the pairwise dissimilarity between the corresponding rows, $\Im_d$ and $\Im'_d$, $1 \le d \le \min\left(\mathfrak{D}(G), \mathfrak{D}(G')\right)$ as

$$W_d = \sqrt{\frac{\sum_{n=1}^{\min(N,N')} (\Im_{dn} - \Im'_{dn})^2}{\min(N, N')}}, \tag{1.29}$$

where N and N' are the sizes and $\mathfrak{D}(G)$ and $\mathfrak{D}(G')$ are the diameters of the spatial graphs G and G', respectively.

While comparing the city spatial graphs of different sizes, we use the "upper left corners" of their cumulative structure matrices bounded by the minimal diameter and the size of the minimal graph. By the way, we suggest that these are close neighbors located just a few steps away from a node which have much stronger impact on the properties of the entire network than those which seem to be statistically isolated from the core of graphs. Therefore, it seems reasonable following Bagrow et al. (2008) to assign weights for the contributions coming from the close neighbors,

$$\omega_d = \sum_{n=1}^{\min(N,N')} (J_{dn} + J'_{dn}), \tag{1.30}$$

and then define the structural distance between two graphs by

$$\Delta\left(G, G'\right) = \frac{\sum_{d=1}^{\min\left(\mathfrak{D}(G), \mathfrak{D}(G')\right)} \omega_d W_d}{\sum_{d=1}^{\min(\mathfrak{D}(G), \mathfrak{D}(G'))} \omega_d}. \tag{1.31}$$

The metric (1.31) quantifying the dissimilarities between the probability distributions of the numbers of the nearest and faraway neighbors in two different graphs equals zero if these distributions are identical (at least on the "upper left corner" of the cumulative structure matrices).

Structural distances between the spatial graphs of six compact urban patterns calculated in accordance with (1.31) are collected in Table 1.2.

It is important to note that the structural distances given in Table 1.2 have been calculated independently for each pair of urban patterns according to their sizes and probability distributions. Clearly, these probability distributions do not belong to the same probability space and, therefore, the structural distances between different pairs of urban patterns cannot be immediately compared. In other words, although the space of graph structures endowed with the Wasserstein distance (1.28) satisfies some of the metric axioms, the triangle inequality does not hold for that and, therefore, it constitutes merely a semimetric space (see Burago et al. 2001).

Table 1.2 The table of structural distances between the spatial graphs of compact urban patterns

	Bielefeld	Rothenburg	Amsterdam	Neubeckum	Venice	Manhattan
Bielefeld	0	0.043	0.321	0.288	0.139	0.396
Rothenburg	0.043	0	0.127	0.269	0.324	0.412
Amsterdam	0.139	0.127	0	0.345	0.316	0.450
Neubeckum	0.288	0.269	0.345	0	0.293	0.432
Venice	0.321	0.324	0.316	0.293	0	0.244
Manhattan	0.396	0.412	0.432	0.450	0.244	0

1.5 Summary

We have discussed the problem of future urbanization and reviewed one of the most conspicuous and robust empirical facts in urban studies, the Zipf Law.

Urban area networks can be abstracted by various graphs accordingly to different paradigms. A graph representing urban space (a spatial graph) is essentially different from the usual city plan. It is fascinating that the construction algorithms for spatial graphs are similar to the basic neurophysiological mechanisms of space apprehension in mammals and birds.

Urban area networks usually possess a very complex structure and may be subjected to the complex network theory analysis.

The measure of integration is one of the principal empirical measures used in space syntax theory to evaluate how many nodes can be reached from given one. The integration probability distribution distinguishes between different urban structures. In particular, it clearly indicates that the partial redevelopments performed in the downtown of Bielefeld during the previous centuries enhanced the effect of ghettoization of the relatively isolated quarters in the ancient part of the city, but, at the same time, reinforced the city core by giving it the crucial role in sustaining the city's connectedness.

We discussed the famous rank-size distribution of city sizes known as Zipf's Law and reported on the similar universality of the Rank-Centrality distributions being fitted perfectly with the linear pattern, with the slope of the line equals to 1, observed for the different urban patterns.

We have also shown that the simple trade-off models can be applied to spatial graphs in order to predict the appearance of urban sprawl with the local land use scheme. In particular, the decisive factor for the emergence of star graphs is the supremacy of the centrality (integration) objective, while the physical (Euclidean) distance between graph vertices is of no importance. It is well-known that the humbleness of physical distances is among the main factors shaping the sprawl land use patterns. We have suggested that the emergence of a complicated highly inhomogeneous structure which we observe in urban developments can generally involve trade-offs, the optimization problems between the multiple, complicated and probably conflicting objectives.

In the last section, we have discussed the automatic classification methods based on the structural statistics of the nearest and faraway neighbors in spatial city networks.

Chapter 2
Wayfinding and Affine Representations of Urban Environments

While travelling in the city, our primary interest is often in finding the best route from one place to another. Since the wayfinding process is a purposive, directed, and motivated activity (Golledge 1999), the shortest route is not necessarily the best one. If an origin and a destination are not directly connected by a continuous path, wayfinding may include search and exploration actions for which it may be crucial to recognize juxtaposed and distant landmarks, to determine turn angles and directions of movement, and eventually to embed the route in some large reference frame.

In Sect. 1.2.2, we have discussed the neurophysiological evidence confirming that the subjective conceptions of environments in animals are formed in the dMEC cortex, one layer upstream the hippocampus, and then stored in hippocampus, in the animal's spatial memory. The topographical regularity and conditional stability of equilateral tessellating triangles of entorhinal grid cells spanning environmental space suggest that physiological interpretation of space in animals is presumably Euclidean, or at least affine and, thus, can be summarized in terms of points, lines, and surfaces.

It is well-known that the conceptual representations of space in humans do not bear a one-to-one correspondence with actual physical space. For instance, in visual representations given by orthographic projection, for a small view, the components of Euclidean transformations such as translations and rotations about the line of sight do not contribute to an understanding of depth, so that the deformations along the line of sight cannot be visually detected, (Normann et al. 1993).

It has also been demonstrated that human perception of structures from motion (Koenderink et al. 1991 and Luong et al. 1994) has limited capabilities to integrate information across more than two views that would result in the inability to recover true Euclidean metric structures. Since stretches along the line of sight should be invisible, the inability to perceive them was in strong support of the affine nature of human space apprehension.

In vision, affine transformations are usually obtained when a planar object is rotated and translated in space, and then projected into the eye via a parallel projection (Pollick et al. 1997a). Affine concepts have been investigated in the analysis of image motion and the perception of three-dimensional structure from motion, (Beusmans 1993, Pollick 1997b), the recognition of planar forms, away from the

Ph. Blanchard, D. Volchenkov, *Mathematical Analysis of Urban Spatial Networks,*
Understanding Complex Systems, DOI 10.1007/978-3-540-87829-2_2,
© Springer-Verlag Berlin Heidelberg 2009

eye, (Wagemans et al. 1994), or in the transformation from visual input to motor output approximating the true Euclidean geometry (Flanders et al. 1992, Pollick et al. 1997a).

In a geometric setting, affine transformations (Zwillinger 1995) are precisely the functions that map straight lines to straight lines, i.e., preserves all linear combination in which the sum of the coefficients is 1. The group of all invertible affine maps of space consists of linear transformations of a point $\mathbf{x}$ into another point $\mathbf{y}$ described by some matrix $\mathscr{A}$ followed by a translation (described by some vector $\mathbf{a}$),

$$\mathbf{y} = \mathscr{A}\mathbf{x} + \mathbf{a}. \tag{2.1}$$

It is known that the affine geometry keeps the concepts of straight lines and parallel lines, but not those of distance between points or value of angles (Busemann et al. 1953). In contrast to (2.1), Euclidean transformations preserve distances between points being a subset of the group of affine transformations such that $\mathscr{A}$ is a special orthogonal matrix (describing rotations), $\mathscr{A}\mathscr{A}^{\top} = 1$, with the transposed matrix $\mathscr{A}^{\top}$.

Being guided primarily by mental representations of locations and neighborhoods, humans interact with physical environments by travelling through them (Bovy et al. 1990) and by communicating within them with other people by chance, (Hillier et al. 1984).

In this chapter, we address the problem of human understanding of spatial configurations. The process of integrating the local affine models of individual places into the entire cognitive map of the urban area network is very complicated and falls largely within the domain of cognitive science and psychology, but nevertheless the specification of what may be recovered from spatial memory can be considered as a problem of mathematics – "the limits of human perception coincide with mathematically plausible solutions" (Pollick 1997b).

In the forthcoming sections, we use discrete time Markov chains (Markov 1971) to construct the affine representations of urban area networks and demonstrate that our algorithm helps to capture a neighborhood's inaccessibility, which could expose hidden islands of future deprivation in cities.

2.1 From Mental Perspectives to the Affine Representation of Space

Human perception of complex spatial structures is always based on the emergence of simplified models that speed up the interpretation process of environments. They may be supported not only by the structure of Euclidean metric space, but also by weaker and, therefore more general structures of affine and projective spaces, (Faugeras 1995). The occasional overlay of geometrically different structures may give rise to multiple visual illusions.

"Illusions of the senses tell us the truth about perception," said Jan Evangelista Purkynje (1787–1869), the first Professor of Physiology at the University of Prague who discovered the Purkinje effect, whereby as light intensity decreases red objects seem to fade faster than blue objects of the same brightness.

Luminance-based repeated asymmetric patterns (RAPs) cause many peoples' visual systems to infer the presence of motion where there is none (Backus et al. 2005). The Japan artist A. Kitaoka has created hundreds of static patterns that appear to move (like that of "rotating snakes" available at $http://www.ritsumei.ac.jp/akitaoka/rotsnake.gif$). The rate of luminance adaptation in human visual cortex is disproportionately faster at high contrast bars coded as deviations from a reference luminance. The tune of the high luminance bars in RAPs by spatial frequency (Backus et al. 2005), the perspective sizes, and the alignment curves (Grzywacz et al. 1991) activates an appropriate set of local velocity detectors in the brain, so that an observer will see the illusory rotation of a disk in "rotating snakes" that can be related by the radius of curvature to motion at constant affine velocity (Pollick 1997b).

The problem with reconstructing of the Euclidean metric properties of space from the affine or projective structures obtained from motion of one or several cameras has always been at the forefront of the interest in computer vision (Koenderink et al. 1991, Faugeras 1995). Affine and projective reconstructions of the Euclidean geometry of the real world have been performed from point correspondences by Sparr (1991), as well as from a number of point correspondences in Luong et al. (1994) and Faugeras (1995).

Travelling through the environment provides spatial knowledge of the city for people, and allows common frames of reference to be established (Golledge 1999). It is suggested in wayfinding studies that the spatial models representing the individual places in each neighborhood can be integrated all along the one-dimensional route trajectories into a single layout covering of the entire city.

Supposing the inherent mobility of humans and alikeness of their spatial perception aptitudes, one might argue that nearly all people experiencing the city would agree in their judgments on the total number of individual locations in that, in identifying the borders of these locations, and their interconnections. In other words, we assume that spatial experience in humans intervening in the city may be organized in the form of a universally acceptable network.

Well-known and frequently travelled path segments provide linear anchors for certain city districts and neighborhoods that help to organize collections of spatial models for the individual locations into a configuration representing the mental image of the entire city (Lynch 1960). In our study, we assume that the frequently travelled routes are nothing else but the " projective invariants" of the given layout of streets and squares in the city – the function of its geometrical configuration, which remains invariant whatever origin-destination route is considered. The arbitrary linear transformations of the geometrical configuration with respect to which a certain property remains invariant constitute the generalized affine transformations.

It is intuitively clear that if the spatial configuration of the city is represented by a regular graph, where each location represented by a vertex has the same number of neighbors, in absence of other local landmarks, all paths would probably be equally followed by travellers. No linear anchors are possible in such an urban pattern which could stimulate spatial apprehension. However, if the spatial graph of the city is far from being regular, then a configurational disparity of different places in the city would result in that some of them may be visited by travellers more often than others.

Random walks provide us with an effective tool for the detailed structural analysis of connected undirected graphs exposing their symmetries (Blanchard et al. 2008).

2.2 Undirected Graphs and Linear Operators Defined on Them

The human minds cogitations about relations between things, beings, and concepts can often be abstracted as a graph that appears to be the natural mathematical tool for facilitating the analysis (Beck et al. 1969, Biggs et al. 1986).

All elements v (nodes or vertices) that fall into one and the same set V are then considered essentially identical, and permutations of them within this set are of no consequence. The symmetric group $\mathbb{S}_N$ consisting of all permutations of N elements (N is the cardinality of the set V), forms the symmetry group of V. If we denote the set of pairwise relationships (edges or links) between all elements of V by $E \subseteq V \times V$, then a graph is a map

$$G(V,E) : E \rightarrow K \subseteq \mathbb{R}_+. \tag{2.2}$$

We emphasize that a graph is an abstract concept, and the definition (2.2) is completely independent of the notions of points and lines which are frequently used to illuminate the structure of the graph. The graph (2.2) is determined by its affinity matrix, $w_{ij} \geq 0$ if $i \sim j$, but $w_{ij} = 0$ otherwise, and is characterized by the set of its automorphisms[1] describing its symmetries. The definitions and notions of graph theory can be found in Biggs (1974), Bollobas (1979) and Diestrel (2005).

A great deal of effort has been devoted by graph theorists to the study of relations between the structure of the graph G and the spectra of different matrices associated to it. The relationship between the graph and the eigenvalues and eigenvectors of its adjacency matrix or of the matrix associated to Laplace operators is the main object of spectral graph theory (see Chung 1989a, 1997, deVerdiere 1998, Godsil et al. 2004 and Chap. 3).

2.2.1 Automorphisms and Linear Functions
         of the Adjacency Matrix

It is well-known (see, for example Biggs 1979, Chap. 4) that a transitive permutation group may be represented graphically, and the converse is also true: a graph

---

[1] An automorphism of a graph is a mapping of vertices such that the resulting graph remains isomorphic to the initial one.

gives rise to a permutation group. An automorphism of a graph G with vertex-set V and edge-set E is a permutation Π of V such that $(i, j) \in E$ implies that $(\Pi(i), \Pi(j)) \in E$. The set of all automorphisms forms a permutation group, $\mathrm{Aut}(G)$, acting on V. For example, the full group of automorphisms of the complete graph $\mathbb{K}_N$ is the symmetric group $\mathbb{S}_N$, since any permutation of the vertices in $\mathbb{K}_N$ preserves adjacency.

In general, $\mathrm{Aut}(G)$ includes all admissible permutations $\Pi \in \mathbb{S}_N$ taking the node $i \in V$ to some other node $\Pi(i) \in V$. The representation of $\mathrm{Aut}(G)$ consists of all $N \times N$ matrices Π_Π, such that $(\Pi_\Pi)_{i,\Pi(i)} = 1$, and $(\Pi_\Pi)_{i,j} = 0$ if $j \neq \Pi(i)$.

A linear transformation of the adjacency matrix

$$Z(\mathbf{A})_{ij} = \sum_{s,l=1}^{N} \mathscr{F}_{ijsl} A_{sl}, \quad \mathscr{F}_{ijsl} \in \mathbb{R}, \tag{2.3}$$

belongs to $\mathrm{Aut}(G)$ if

$$\Pi_\Pi^\top Z(\mathbf{A}) \Pi_\Pi = Z\left(\Pi_\Pi^\top \mathbf{A} \Pi_\Pi\right), \tag{2.4}$$

for any $\Pi \in \mathrm{Aut}(G)$. It is clear that the relation (2.4) is satisfied if the entries of the tensor $\mathscr{F}$ in (2.3) meet the following symmetry property:

$$\mathscr{F}_{\Pi(i)\Pi(j)\Pi(s)\Pi(l)} = \mathscr{F}_{ijsl}, \tag{2.5}$$

for any $\Pi \in \mathrm{Aut}(G)$. Since the action of the symmetry group preserves the conjugate classes of index partition structures, it follows that any appropriate tensor $\mathscr{F}$ satisfying (2.5) can be expressed as a linear combination of the following tensors:

$$\begin{aligned} \{ 1, \delta_{ij}, \delta_{is}, \delta_{il}, \delta_{js}, \delta_{jl}, \delta_{sl}, \delta_{ij}\delta_{js}, \delta_{js}\delta_{sl}, \\ \delta_{sl}\delta_{li}, \delta_{li}\delta_{ij}, \delta_{ij}\delta_{sl}, \delta_{is}\delta_{jl}, \delta_{il}\delta_{js}, \delta_{ij}\delta_{il}\delta_{is} \} \, . \end{aligned} \tag{2.6}$$

By substituting the above tensors into (2.3) and taking the symmetries, into account we find that any arbitrary linear permutation invariant function $Z(\mathbf{A})$ defined on a simple undirected graph $G(V,E)$ must be of the following form,

$$Z(\mathbf{A})_{ij} = a_1 + \delta_{ij}(a_2 + a_3 k_j) + a_4 A_{ij}, \tag{2.7}$$

where $k_j = \deg(j)$, and $a_{1,2,3,4}$ being arbitrary constants.

If we impose, in addition, that the linear function Z preserves the connectivity,

$$\begin{aligned} \Sigma_{i \sim j} A_{ij} &= \\ \deg(i) &\equiv k_i \\ &= \Sigma_{j \in V} Z(\mathbf{A})_{ij}, \end{aligned} \tag{2.8}$$

then it follows that $a_1 = a_2 = 0$ (since the contributions of $a_1 N$ and a_2 are indeed incompatible with (2.8)), and the remaining constants should satisfy the relation $1 - a_3 = a_4$. By introducing the new parameter $\beta \equiv a_4 > 0$, we can reformulate (2.7) in the following form,

$$Z(\mathbf{A})_{ij} = (1 - \beta)\,\delta_{ij}k_j + \beta A_{ij}. \tag{2.9}$$

It is important to note that (2.8) can be interpreted as a probability conservation relation,

$$1 = \frac{1}{k_i}\sum_{j \in V} Z(\mathbf{A})_{ij}, \quad \forall i \in V, \tag{2.10}$$

and, therefore, the linear function, $Z(\mathbf{A})$ can be associated to a stochastic process (Blanchard et al. 2008).

By substituting (2.9) into (2.10), we obtain

$$\begin{aligned} 1 &= \sum_{j \in V}(1 - \beta)\,\delta_{ij} + \beta\frac{A_{ij}}{k_i} \\ &= \sum_{j \in V} T_{ij}^{(\beta)}, \end{aligned} \tag{2.11}$$

in which the operator

$$T_{ij}^{(\beta)} = (1 - \beta)\,\delta_{ij} + \beta\frac{A_{ij}}{k_i} \tag{2.12}$$

is nothing else but the transition operator of a generalized random walk, for $\beta \in [0,1]$. The operator $T_{ij}^{(\beta)}$ defines "lazy" random walks for which a random walker stays in the initial vertex with probability $1 - \beta$, while it moves to another node randomly chosen among the nearest neighbors with probability β/k_i. In particular, for $\beta = 1$, the operator $T_{ij}^{(\beta)}$ describes the standard random walks extensively studied in the classical surveys of Lovasz (1993) and Aldous et al. 2008 inprepration

However, if instead of the probability conservation relation (2.10) we require that the linear function $Z(\mathbf{A})$ is harmonic, i.e.,

$$\sum_{j \in V} Z(\mathbf{A})_{ij} = 0, \quad \forall i \in V, \tag{2.13}$$

then it has been proven by Smola et al. (2003) that (2.9) recovers the generalized Laplace operator

$$L_{ij} = -\frac{a_2}{N} + \delta_{ij}\left(a_2 + a_3\deg(i)\right) - a_3 A_{ij}, \tag{2.14}$$

which describes the diffusion processes characterized by the conservation of mass.

The choice of a_2 and a_3 depends upon the details of the transport process in question. The constant a_2 determines a zero-level transport mode and is usually taken as $a_2 = 0$. The Laplace operator (2.14) where $a_2 = 0$ and $a_3 = 1$ is called the canonical Laplace operator, (deVerdiere 1998),

$$\mathbf{L_c} = \mathbf{D} - \mathbf{A}, \tag{2.15}$$

where $\mathbf{D}$ is the diagonal matrix, $\mathbf{D} = \mathrm{diag}\left(\deg(1),\ldots\deg(N)\right)$.

We conclude that the random walk transition operator and the Laplace operator are nothing else but the representations of the set of automorphisms of the graph in the classes of stochastic and harmonic matrices, respectively (Blanchard et al. 2008).

It is also important to mention that the transition probability operator (2.12) describing the set of paths available from $i \in V$ constitutes the probabilistic analog of the affine transformations remaining invariant the probability distribution π, namely the stationary distribution of random walks.

2.2.2 Measures and Dirichlet Forms

The nodes of the graph $G(V,E)$ can have different weights (or masses) accounted by some measure

$$m = \sum_{i \in V} m_i \, \delta_i \qquad (2.16)$$

specified by any set of positive numbers $m_i > 0$. For example, the counting measure assigns to every node a unit mass,

$$m_0 = \sum_{i \in V} \delta_i. \qquad (2.17)$$

The Hilbert space $\mathscr{H}$ (a complete inner product space of functions) on $\mathbb{R}^V$ is the space of squared summable real-valued functions $\ell^2(m_0)$ endowed with the usual inner product

$$\langle f \,|\, g \rangle = \sum_{i \in V} f(i)\,g(i),$$

for all $f, g \in \mathscr{H}(V)$.
  Vectors

$$\mathbf{e}_i = (0, \ldots, 1_i, \ldots 0)$$

with a unit at the ith position representing the node $i \in V$ form the canonical basis in Hilbert space $\mathscr{H}(V)$.
  For undirected graphs, the Hilbert space structure on $\mathbb{R}^E$ is represented by a symmetric Dirichlet form on $f \in \ell^2(m_0)$ over all edges $i \sim j$,

$$\mathscr{D}(f) = -\sum_{i \sim j} c_{ij} \left(f(i) - f(j) \right)^2, \qquad (2.18)$$

in which $c_{ij} = c_{ji} \geq 0$. The quadratic form (2.18) is associated to the elliptic canonical Laplace operator Δ_G,

$$\mathscr{D}(f) = \langle f, \Delta_G f \rangle, \quad f \in \ell^2(m_0), \qquad (2.19)$$

which is self-adjoint with respect to the counting measure m_0.
  The canonical Laplace operator can be represented as a product

$$\Delta_G = d^\top d = d\,d^\top \qquad (2.20)$$

of the difference operator $d : \mathbb{R}^V \to \mathbb{R}^E$,

$$d_{ij}(f) = \begin{cases} f(i) - f(j), & i \sim j, \\ 0, & \text{otherwise}, \end{cases} \tag{2.21}$$

and its adjoint $d^\top$.

It is remarkable that the measure m_0 is not the unique measure that can be defined on V. Given a set of real positive numbers $m_j > 0$ other measures can be defined by

$$m = \sum_{j \in V} m_j \delta_j. \tag{2.22}$$

The measure associated with random walks defined on undirected graphs,

$$m = \sum_{j \in V} \deg(j) \delta_j, \tag{2.23}$$

is an example.

The transition to the new measure implies a suitable transformation of functions

$$R_m : f_m(j) \rightarrow m_j^{-1/2} f(j), \quad j \in V, \tag{2.24}$$

preserving the notion of elliptic differential operators defined on $\mathbb{R}^V$. The Laplace operator self-adjoint with respect to the measure m is unitary equivalent to Δ_G,

$$L_m = R_m^{-1} \Delta_G R_m, \tag{2.25}$$

where R_m is the transformation (2.24).

Similarly, the random walks transition operator self-adjoint with respect to the measure m is unitary equivalent to that of T^β,

$$T_m = R_m^{-1} T R_m. \tag{2.26}$$

The matrices associated with unitary equivalent operators share many properties: they have the same rank, the same determinant, the same trace, the same eigenvalues (though the eigenvectors will, in general, be different), the same characteristic polynomial and the same minimal polynomial. These can, therefore, serve as isomorphism invariants of graphs. However, two graphs may possess the same set of eigenvalues, but not be isomorphic (deVerdiere 1998).

2.3 Random Walks Defined on Undirected Graphs

In the present section, we consider a transition operator determining time reversible random walks of the nearest neighbor type $(V, \mathbf{T})$ where V is the vertex set of $G(V, E)$ and

$$
\begin{aligned}
T_{ij} &= \Pr\left[v_{t+1} = j \,|\, v_t = i\right] > 0 \Leftrightarrow i \sim j, \\
&= \mathbf{D}^{-1}\mathbf{A} \\
&= \deg(i)^{-1}, \text{ iff } i \sim j,
\end{aligned}
\tag{2.27}
$$

is the one-step probability of a Markov chain $\{v_t\}_{t \in \mathbb{N}}$ (see Markov) with state space V (states can repeat). The discrete time random walks introduced on graphs have been studied in Lovasz (1993), Lovasz et al. (1995) and Saloff-Coste (1997).

2.3.1 Graphs as Discrete time Dynamical Systems

A finite graph $G(V,E)$ can be interpreted as a discrete time dynamical system with a finite number of states (Prisner 1995). The temporal evolution of such a dynamical system is described by a "dynamical law" that maps vertices of the graph into other vertices. The Markov transition operator (2.27) is related to the unique Perron-Frobenius operator of the dynamical system, (Mackey 1991).

We can consider a transformation $\mathscr{S} : V \to V$ such that it maps any subset of nodes $U \subset V$ into the set of their neighbors,

$$
\mathscr{S}(U) = \{w \in V \,|\, v \in U, v \sim w\}.
\tag{2.28}
$$

We denote the result of $t \geq 1$ successive applications of the transformation to $U \subset V$ as $\mathscr{S}_t(U)$. Then, for every vertex $v \in V$, the sequence of successive points $\mathscr{S}_t(v)$ considered as a function of time is called a trajectory. Given a density function $f(v) \geq 0$ such that $\sum_{v \in V} f(v) = 1$, it follows from (2.27) that the dynamics are given by

$$
f^{(t+1)} = f^{(t)} T.
\tag{2.29}
$$

By definition, the operator T^t is the Perron-Frobenius operator corresponding to the transformation $\mathscr{S}$, since

$$
\sum_{v \in U} f(v) \, T^t = \sum_{\mathscr{S}_t^{-1}(U)} f(v).
\tag{2.30}
$$

The uniqueness of the Perron-Frobenius operator T^t is a consequence of the Radon-Nikodym theorem (Shilov et al. 1978).

2.3.2 Transition Probabilities and Generating Functions

Given a random walk $(V, \mathbf{T})$, we denote the probability of transition from i to j in $t > 0$ steps by

$$
p_{ij}^{(t)} = \left(\mathbf{T}^t\right)_{ij}.
\tag{2.31}
$$

The generating function (the Green function) of the transition probability (2.31) is defined by the following power series:

$$G_{ij}(z) = \sum_{t \geq 0} p_{ij}^{(t)} z^t$$
$$= (\mathbf{1} - z\mathbf{T})^{-1},$$
(2.32)

which is convergent inside the unit circle $|z| < 1$, the spectral radius of the positive stochastic matrix $\mathbf{T}$. Then, it can be readily demonstrated that the limit

$$\mathbf{T}^{\infty} = \lim_{t \to \infty} \mathbf{T}^t$$
(2.33)

exists as a positive stochastic matrix.

The first hitting probabilities,

$$q_{ij}^{(t)} = \Pr[v_t = j, v_l \neq j, l \neq 1, \ldots, t-1 | v_0 = i], \quad q_{ij}^{(0)} = 0,$$
(2.34)

are related to the transition probability $p_{ij}^{(t)}$ by

$$p_{ij}^{(t)} = \sum_{s=0}^{t} q_{ij}^{(s)} p_{jj}^{(t-s)}$$
(2.35)

and can be calculated by means of the generating function

$$F_{ij}(z) = \sum_{t \geq 0} q_{ij}^{(t)} z^t, \quad i, j \in V, \quad z \in \mathbb{C}.$$
(2.36)

It follows from (2.34) that the generating functions (2.32) and (2.36) are related (Lovasz et al. 1995) by the equation

$$G_{ij}(z) = F_{ij}(z) G_{jj}(z)$$
(2.37)

and, therefore, $F_{ij}(z)$ is nothing else as the renormalized Green function $G_{ij}(z)$ in such a way that its diagonal entries become 1.

2.3.3 Stationary Distribution of Random Walks

For the operator T defined on a connected aperiodic graph G, the Perron–Frobenius theorem see (Graham 1987, Minc 1988, Horn et al. 1990) asserts that its largest eigenvalue $\mu_1 = 1$ is simple and the eigenvector belonging to it is strictly positive.

A fundamental result on random walks introduced on undirected graphs (Lovasz 1993, Lovasz et al. 1995) is that among all possible distributions

$$\sigma_i = \Pr[v_t = i]$$
(2.38)

defined on the graph there exists a *unique* stationary distribution $\pi : V \rightarrow [0,1]^N$, solution of the eigenvalue problem,

$$\pi \mathbf{T} = 1 \pi, \tag{2.39}$$

satisfying the detailed balance equation (Aldous et al. in prepration),

$$\pi_i T_{ij} = \pi_j T_{ji}, \tag{2.40}$$

from which it follows that a random walk considered backwards is also a random walk (time reversibility property). For the nearest neighbor random walk defined by (2.27), the stationary distribution equals

$$\pi_i = \frac{k_i}{2M}, \quad \sum_{i \in V} \pi_i = 1 \tag{2.41}$$

where M is the total number of edges in the graph, and k_i is the degree of the node i.

Given the stationary distribution of random walks, we then define the symmetric transition matrix by

$$\widehat{T}_{ij} = \pi_i^{1/2} T_{ij} \pi^{-1/2} \tag{2.42}$$

and transform it to a diagonal form,

$$\widehat{\mathbf{T}} = U \Lambda U^T, \tag{2.43}$$

where U is an orthonormal matrix, and Λ is a real diagonal matrix. The symmetric transition matrix for the random walk defined on undirected graphs is simply

$$\widehat{T}_{ij} = \frac{A_{ij}}{\sqrt{k_i k_j}}. \tag{2.44}$$

The diagonal entries of Λ,

$$1 = \mu_1 > \mu_2 \geq \ldots \geq \mu_N \geq -1, \tag{2.45}$$

are the eigenvalues of the transition matrices $\widehat{\mathbf{T}}$ and $\mathbf{T}$. These eigenvalues correspond to the left and right eigenvectors,

$$x_i = \sum_{j \in V} \alpha_j \sqrt{\pi_i} \, U_{ij}, \quad y_i = \sum_{j \in V} \frac{\alpha_j'}{\sqrt{\pi_i}} \, U_{ij}, \tag{2.46}$$

(α_j and α_j' are arbitrary constants) of the transition matrix:

$$\sum_{i \in V} x_i T_{ij} = \mu x_j, \quad \sum_{i \in V} T_{ij} y_i = \mu y_j, \quad \forall j \in V. \tag{2.47}$$

The symmetric matrix $\widehat{\mathbf{T}}$ as well as its powers can then be written using the spectral theorem,

$$\widehat{\mathbf{T}} = \sum_{i=1}^{N} \mu_i |x_i\rangle \langle y_i|, \quad \widehat{\mathbf{T}^n} = \sum_{i=1}^{N} \mu_i^n |x_i\rangle \langle y_i|. \tag{2.48}$$

It follows from (2.48) that the transition probability $p_{ij}^{(t)}$ can be computed in the following way,

$$p_{ij}^t = \pi_j + \sum_{l=2}^{N} \mu_l^t x_{il} y_{jl}. \tag{2.49}$$

2.3.4 Continuous Time Markov Jump Process

We can consider a continuous time Markov jump process $\{w_t\}_{t \in \mathbb{R}_+} = \{v_{Po(t)}\}$ where $Po(t)$ is the Poisson distribution instead of the discrete time Markov chain $\{v_t\}_{t \in \mathbb{N}}$ (Aldous et al. 2008 in prepration). Supposing that the transition time τ is a discrete random variable distributed with respect to the Poisson distribution $Po(t)$, we can write down the corresponding operator with mean 1 exponential holding times as

$$\begin{aligned} p_{ij}^t &= \pi_j + \sum_{l=2}^{N} x_{il} y_{jl} \sum_{\tau=0}^{\infty} \mu_l^\tau \frac{t^\tau e^{-t}}{\tau!} \\ &= \pi_j + \sum_{l=2}^{N} x_{il} y_{jl} e^{-t\lambda_l}, \end{aligned} \tag{2.50}$$

where $\lambda_l \equiv (1 - \mu_l)$ is the lth *spectral gap*. The relaxation processes towards the stationary distribution π of random walks are described by the characteristic decay times $\tau_l = -1/\ln \lambda_l$. The asymptotic rate of convergence for (2.50) to the stationary distribution is determined by the spectral gap $\lambda_2 = 1 - \mu_2$.

The stationary distribution for general directed graphs is not so easy to describe; it can be very far from a uniform one since the probability that some nodes could be visited may be exponentially small in the number of edges (Lovasz et al. 1995). Moreover, if a directed graph has cycles such that the common divisor of their lengths is larger than 1, the random walk process defined on it does not have any stationary distribution.

2.4 Study of City Spatial Graphs by Random Walks

The issues of global connectivity of finite graphs and accessibility of their nodes have always been classical fields of research in graph theory. The level of accessibility of nodes and subgraphs of undirected graphs can be estimated precisely in connection with random walks introduced on them (Volchenkov et al. 2007a, Rosvall et al. 2008). Although random walkers do not interact with each other, the statistical properties of their flows could be highly nontrivial being a detailed fingerprint of the topology. In the present section, we discuss how to analyze and measure the structural dissimilarity between different locations in complex urban networks by means of random walks.

2.4.1 Alice and Bob Exploring Cities

We explore the spatial graphs of urban environments following two different strategies of discrete time random walks personified by two walkers, A (Alice) and B (Bob), respectively. Alice and Bob start walking from a location $x_0 \in V$ randomly chosen among all available locations in the city.

Alice moves at each time step from its actual node x_t to the next one, $x_{t+1} \neq x_t$, selecting it randomly among all other locations adjacent to x_t, so that Alice's walks constitute a discrete time Markov chain, $\mathscr{X} = [X_0, X_1, \ldots X_t], t \in \mathbb{N}$, where

$$\Pr\left(X_{t+1} = x_{t+1} | X_t = x_t, \ldots X_1 = x_1\right) = \Pr\left(X_{t+1} = x_{t+1} | X_t = x_t\right),$$

for all $x_0, x_1, \ldots x_{t+1} \in V$. The Markov chain $\mathscr{X}$ is determined by its initial site x_0 and the probability transition matrix between the adjacent sites $x_i \sim x_j$,

$$T_{x_i, x_j}^{(A)} = \frac{A_{x_i, x_j}}{\deg\left(x_i\right)} \tag{2.51}$$

where A_{x_i, x_j} is the entry of the $\{0, 1\}$ adjacency matrix of the city spatial graph, and $\deg\left(x_i\right)$ is the number of neighboring places x_i is adjacent to. Since the city spatial graph is assumed to be connected and undirected, it is possible to go by $\mathscr{X}$ with positive probability from any city location to any other one in a finite number of steps, so that the Markov chain $\mathscr{X}$ is irreducible and time reversible.

The random walk of Bob, $\mathscr{Y} = [Y_0, Y_1, \ldots Y_t], t \in \mathbb{N}$, is biased in favor of nodes with the high centrality index,

$$T_{y_i, y_j}^{(B)} = \frac{m_{y_j} A_{y_i, y_j}}{\sum_{y_s \in V} m_{y_s} A_{y_i, y_s}}, \tag{2.52}$$

where m_{y_j} is the total number of shortest paths between all pairs of distinct locations in the spatial graph of the city that pass through the place y_j. The random walk $\mathscr{Y}$ also constitutes a time reversible irreducible Markov chain. However, it is clear that among all places adjacent to his current location Bob always prefers to move into those which occur on the shortest paths in the city graph. In other words, with higher probability, Bob chooses those places that are characterized by the higher betweenness centrality index (i.e., being of a strong choice, in the space syntax terminology) than those that do not. We must stress that the transition operator suggested in (2.52) does not imply that a place that is not a strong choice would never be visited by Bob; however, such a visit is less probable.

Random walks defined by (2.52) are related to various practical studies concerning the city and intercity routing problems, such as the travelling salesman problem, in which the cheapest route is searched (Dantzig et al. 1954). They are also related to pedestrian surveys performed in the framework of space syntax research that offers evidence that people, in general, prefer to move through the more central (integrated) places in the city (Hillier 2004).

The crucial difference between these two strategies is that while the transition operator (2.51) respects the structure of the graph as captured by its automorphism group being a particular case of the operator (2.12) for $\beta = 1$, Bob's shortest path strategy defined by (2.52) does not.

Indeed, the "shortest path strategy" represented above is only one among infinitely many other strategies that walkers – whether they are random or not – would follow while surfing through the city. Given a set of positive masses $m_i > 0$ characterizing the attraction of a particular place in the city, one can define the correspondent biased walk by (2.52). However, none of them actually fit the set of graph automorphisms, except one of Alice's, with $m_i = 1$.

2.4.2 Mixing Rates in Urban Sprawl and Hell's Kitchens

At the onset of random walks, many new places are visited for the first time and then revisited again until the variations of visiting frequency decreases substantially. Later on, when discovering new nodes takes more time, the rate at which these variations decrease becomes even slower, until the stationary distribution of random walks is eventually achieved.

The rate of convergence,

$$\eta = \lim_{t \to \infty} \sup \max_{i,j} \left| p_{ij}^{(t)} - \pi_j \right|^{1/t}, \tag{2.53}$$

called the mixing rate (Lovasz et al. 1995) is a measure of how fast the stationary distribution of random walks π can be achieved on the given graph G. The mixing time,

$$\tau = -\frac{1}{\ln \eta}, \tag{2.54}$$

being a reciprocal quantity for (2.53) measures the expected number of steps required to achieve the stationary distribution of random walks.

If defined for the city spatial graph, the mixing rate (2.53) can be used as a quantitative measure of its structural regularity. If calculated for city spatial graphs, its value is close to 1 if the spatial structure of the urban pattern contains a great deal of repeating elements, but it is below 1 if the city is less ordered. It is known from the space syntax research (see Jiang 1998) that the level of regularity of urban environments is among the key factors that determines people's orientation perception and their wayfinding abilities.

In the Chapter 1, we discussed how similar geometrical elements found in the urban pattern in repetition are converted into a set of twin nodes in the city spatial graph. In particular, an urban pattern developed in an ideal grid is represented by the complete bipartite graph since, for all twin nodes, the rows and columns of the correspondent adjacency matrix are identical. Twin nodes in the graph have no consequences for random walks.

A spatial structure of suburban sprawl is represented by a star graph, in which individual spaces of the private parking places are connected to a hub associated with the only sinuous central road. Cliental nodes (of degree 1) of a star graph are also twins, being connected to the same hub at the center.

Provided the spatial graph G contains $2n$ twin nodes, the correspondent transition probability matrix (2.44) has the $2n-2$ multiple eigenvalue $\mu = 0$. In particular, the spectrum of a star graph consists of two simple eigenvalues, 1 and -1, and $2n-2$ eigenvalues $\mu = 0$. The linear vector subspace which belongs to the multiple eigenvalue $\mu = 0$ is spanned by $2n-2$ orthonormal Faria vectors,

$$\hat{f}_s = \frac{1}{\sqrt{2}}[0,\ldots,1,\ldots,-1,\ldots 0], \quad s = 1,\ldots 2n-2,$$

distinguished by the different positions of 1 and -1. It is clear from (2.49) that, due to $\mu = 0$, none of them contributes to the mixing rate (2.53). The transition probability between them is independent of time, $p_{ij}|_{i,j-\text{twins}} = 1/2(n-1)$ and can be very small if the spatial representation of the urban pattern contains many twin nodes, $n \gg 1$.

In Fig. 2.1, we have represented the comparative diagram for mixing rates of random walks (2.51) defined on the spatial graphs of five compact urban patterns. In order to compare the mixing rates calculated for the organic cities with those found in the urban patterns developed in grids, we have considered a particular neighborhood in Manhattan, in Midtown West, called Hell's Kitchen (also known as

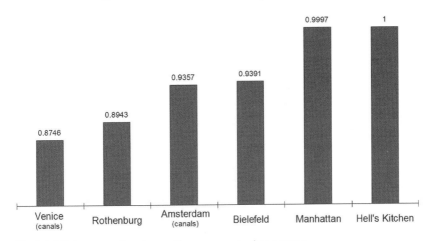

Fig. 2.1 Mixing rates of random walks on compact urban patterns.

Clinton). This neighborhood includes the area between 34th Street and 57th Street, from 8th Avenue to the Hudson River (the famous selling of the musical "West Side Story," written by A. Laurents). For us it is essentially interesting since its spatial structure constitutes an ideal grid formed by the standard blocks of 264×900 square feet.

The values of the mixing rate allows us to order the compact urban patterns with respect to regularity of their spatial structures – from the city canal network in Venice to the neighborhood in Manhattan. The mixing rate of random walks in Hell's Kitchen always equals 1!

2.4.3 Recurrence Time to a Place in the City

The wind blows to the south, and goes round to the north; round and round goes the wind, and on its circuits the wind returns.

(Ecclesiastes 1:6)

Beyond any doubt, thanks to connectedness and equi-directedness of the spatial network the edict can work.

The recurrence time of a location indicates how long a random walker must wait to revisit the site. It is known from the work of (Kac 1947) that, for a stationary, discrete-valued stochastic process, the expected recurrence time to return to a state is just the reciprocal of the probability of this state.

The stationary distribution of Alice's random walk defined by (2.51) is $\pi_i^{(A)} = k_i/2M$. Interestingly, it does not depend on the size of the spatial graph N, but on the total number of edges, M. Consequently, the recurrence time to a location in the random walk of Alice is inversely proportional to connectivity of the place,

$$r_i^{(A)} = \frac{2M}{k_i} \tag{2.55}$$

and, therefore, depends upon the local property of the place (connectivity).

The stationary distribution of Bob's random walk defined by (2.52) is the betweenness centrality index of a place (i.e., its global choice) defined by (1.9), $\pi_i^{(B)} = \text{Choice}(i)$ and, therefore, the recurrence time to a node in Bob's walk given by

$$r_i^{(B)} = \frac{1}{\text{Choice}(i)} \tag{2.56}$$

can be very different from that in Alice's walk. The recurrence time (2.56) depends upon the global property of the place in the city.

The key observation related to the stationary distributions of random walks defined on the city spatial graph is that sometimes a highly connected node (a hub) can have a surprisingly low betweenness centrality and vice versa – the local and global properties of nodes are not always positively correlated. In fact, an urban pattern represented by its spatial graph can be characterized by a certain discrepancy between connectivity and centrality of locations. Such a part-whole relationship between local and global properties of the spaces of motion is known in space syntax theory as intelligibility of urban pattern (Hillier et al. 1984, Hillier 1996). The adequate level of intelligibility is proven to be a key determinant of human behavior in urban environments encouraging people's wayfinding abilities (Jiang et al. 2004).

In order to measure the uncertainty between the connectivity and betweenness centralities of places in the city, we can use the standard Kullback-Leibler distance (the relative entropy between two stationary distributions of random walks, $\pi^{(A)}$ and $\pi^{(B)}$) (Cover et al. 1991),

$$D\left(\pi^{(B)} \middle| \pi^{(A)}\right) = \sum_{i \in V} \pi_i^{(A)} \log \frac{\pi_i^{(A)}}{\pi_i^{(B)}}. \tag{2.57}$$

The relative entropy (2.57) is always nonnegative and is zero if and only if both probability distributions are equal. However (2.57) does not satisfy the triangle inequality and is not symmetric, so that it is not a true distance.

In Fig. 2.2, we have represented the comparative diagram of the relative entropies between stationary distributions of random walks performed by Alice and Bob in five city spatial graphs. The outstandingly high discrepancy between connectivity and centrality of canals in the city canal network of Venice draws the attention.

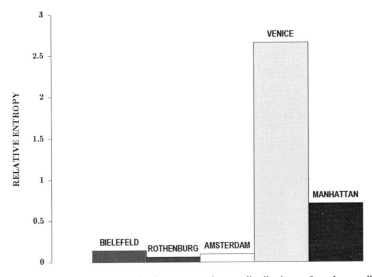

Fig. 2.2 The relative entropies between stationary distributions of random walks performed by Alice and Bob in five city spatial graphs

2.4.4 What does the Physical Dimension of Urban Space Equal?

In classical physics, a traveller has three basic directions in which he or she can move from a particular point – they are physical dimensions of our space.

While simulating the diffusion equation $u_t = \triangle u$ for the scalar function u defined on a regular d-dimensional lattice, $\mathscr{L}_a = a\mathbb{Z}^d$, with the lattice scale length a, one uses its discrete representation,

$$u^{t+1}(x) = \frac{1}{2^d a^2} \left[\sum_{y \in U_x} u^t(y) - 2^d u^t(x) \right],\tag{2.58}$$

where U_x is the lattice neighborhood of $x \in \mathscr{L}_a$. The cardinal number 2^d is uniform for the given lattice and, therefore, the parameter d in (2.58) is interpreted as the dimension of Euclidean space.

Being defined on an arbitrary connected graph $G(V,E)$ the discrete Laplace operator actually has the same form as in (2.58), except when the cardinality number changes to 2^{δ_x} where

$$\delta_x = \log_2 k_x, \quad k_x = \deg(x)\tag{2.59}$$

it may be considered as the local analog of the physical dimension at the node $x \in V$ (Volchenkov et al. 2007c).

An interesting question arises concerning (2.59), namely it is possible to define a global dimensional property of the graph that can be considered as the dimension of space? Below, we show that this can be done on a statistical ground, by estimating the spreading of a set of independent random walkers. In information theory (Cover et al. 1991), such a spreading is measured by means of the entropy rate, the informational analog of the physical dimension of space.

Random walks performed by Alice and Bob on undirected spatial graphs are both Markov chains. The number of all possible paths on the graphs grows exponentially with the length of paths n. Therefore the probability to observe a long enough typical random path $\{X_1 = x_1, \ldots X_n = x_n\}$ decreases asymptotically exponentially,

$$2^{-n(H(\mathscr{X})+\varepsilon)} \leq \Pr[\{X_1 = x_1, \ldots X_n = x_n\}] \leq 2^{-n(H(\mathscr{X})-\varepsilon)}\tag{2.60}$$

where the parameter $nH(\mathscr{X})$ measuring the uncertainty of paths in random walks (*entropy*) grows asymptotically linear with n at a rate $H(\mathscr{X})$ which is, therefore, called the entropy rate. Since random walks performed by Alice and Bob on the undirected spatial graphs are both irreducible and aperiodic, the corresponding entropy rates are independent of the initial distribution, the probability to chose a certain location in the city as a starting point for a walk. Indeed, the value of $H(\mathscr{X})$ depends upon the strategy of random walks and, in general, $H_A \neq H_B$ for any Markov chain $\mathscr{X}$ (Cover et al. 1991):

$$H_{A,B} = - \sum_{x_i, x_j \in V} \pi^{(A,B)} T_{x_i x_j}^{(A,B)} \log_2 \left(T_{x_i x_j}^{(A,B)} \right).\tag{2.61}$$

As usual, in (2.61) we assume that $0 \cdot \log(0) = 0$. In information theory, the entropy rate (2.61) is important as a measure of the average message size required to describe a stationary ergodic process (Cover et al. 1991); provided Alice and Bob use the binary code in their routing reports, they need approximately $nH_A(\mathscr{X})$ and $nH_B(\mathscr{X})$ bits, respectively, in order to describe the typical path of length n. The entropy rates have been used in Boccaletti et al. (2006) and Gomez-Gardenes et al. (2007) as a measure to characterize properties of the topology of complex networks.

Substituting the transition matrix elements and the stationary distribution of Alice's random walks into (2.61), we obtain the entropy rate of an unbiased random walk on a connected undirected network as:

$$H_A = \frac{1}{2M} \sum_{x_i \in V} \delta_x \qquad (2.62)$$

where δ_x is the local analog of space dimensions defined in (2.59). By the way, the entropy rate of Alice's random walks is just an average of $\log_2(k_x)$ over all edges in the graph. The entropy rate of Bob's random walks has no simple expression, but can be readily computed numerically – typically, its value exceeds the entropy rate reported by Alice. In Fig. 2.3, we have presented the comparative diagram of entropy rates for both random walks in all five compact urban patterns.

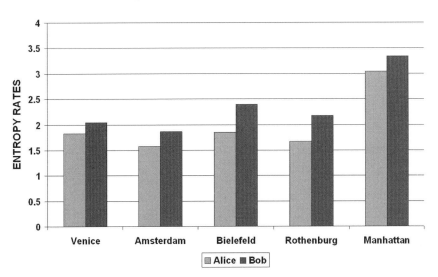

Fig. 2.3 The entropy rates of random walks reported by Bob and Alice for all five compact urban patterns

Interestingly, the dimensions of random walks defined on spatial graphs of cities which are either organic or had experienced the organic phase in their developments are close to 2, so that their urban space is almost planar. In contrast to them, the relatively regular urban street pattern in Manhattan apparently forms a three-dimensional space.

2.5 First-Passage Times: How Random Walks Embed Graphs into Euclidean Space

Discovering important nodes and quantifying differences between them in a graph is not easy since the graph, in general, does not possess the structure of Euclidean space. Representing graph nodes by means of vectors of the canonical basis does not give a meaningful description of the graph since all nodes obviously are equivalent. A natural idea consists of the use of eigenvalues and eigenvectors of a self-adjoint operator defined on the graph in order to define a metric relevant to its topological structure. Spectral methods are popular in applications because they allow a lot of information about graphs to be extracted with minimal computational efforts. The use of self-adjoint operators has become standard in spectral graph theory (Chung 1997), as well as in theory of random walks (Lovasz 1993, Aldous et al. 2008).

2.5.1 Probabilistic Projective Geometry

The stationary distribution of random walks π defines a unique measure on the set of nodes V with respect to which the transition operator ((2.11) for $\beta = 1$) is self-adjoint,

$$\widehat{T} = \frac{1}{2}\left(\pi^{1/2}T\pi^{-1/2} + \pi^{-1/2}T^{\top}\pi^{1/2}\right), \qquad (2.63)$$

where $T^{\top}$ is the adjoint operator, and π is defined as the diagonal matrix diag $(\pi_1, \ldots, \pi_N)$. In particular, for a simple undirected graph, the symmetric operator is defined by (2.44). While interesting in the spectral calculations of random walks characteristics, the symmetric matrix (2.63) is more convenient since its eigenvalues are real and bounded in the interval $\mu \in [-1, 1]$ and the eigenvectors define an orthonormal basis.

An affine coordinate system on V is prescribed by an independent set of points for which the displacement vectors $\mathbf{e}_j = j - i$, $j \in V$, $j \neq i$, form a basis of V with respect to the point $i \in V$. A displacement vector $\mathbf{v} = \sum_{j \in V} v_j \mathbf{e}_j$ is identified with the coordinate $(N-1)$-tuple $(v_1, \ldots, \{ \}_i, \ldots, v_N)$, in which the ith component is missing. We can associate all points V with their relative displacement vectors.

The stationary probability distribution π associated with random walks allows us to define a coordinate system in the projective probability space.

Given a symmetric matrix $w_{ij} \geq 0$ and a vector $\beta_i \in [0,1]$, we can define the transition probability by the kernel (2.27) on V and its self-adjoint counterpart (2.63). The complete set of real eigenvectors $\Psi = \{\psi_1, \psi_2, \ldots \psi_N\}$ of the symmetric matrix (2.63),

$$|\psi_i\rangle \widehat{T} = \mu_i |\psi_i\rangle,$$

ordered in accordance to their eigenvalues, $\mu_1 = 1 > \mu_2 \geq \ldots \mu_N \geq -1$, forms an orthonormal basis in $\mathbb{R}^N$,

$$\langle \psi_i | \psi_j \rangle = \delta_{ij}, \tag{2.64}$$

associated to the linear automorphisms of the affinity matrix w_{ij}, (Blanchard et al. 2008). In (2.64), we have used Dirac's bra-ket notations especially convenient for working with inner products and rank-one operators in Hilbert space.

Given the random walk defined by the operator (2.27), then the squared components of the eigenvectors ψ have very clear probabilistic interpretations. The first eigenvector ψ_1 belonging to the largest eigenvalue $\mu_1 = 1$ satisfies $\psi_{1,i}^2 = \pi_i$ and describes the probability to find a random walker in $i \in V$. The norm in the orthogonal complement of ψ_1, $\sum_{s=2}^N \psi_{s,i}^2 = 1 - \pi_i$, is nothing else, but the probability that a random walker is not in i.

Looking back it is easy to see that the transition operator (2.63) defines a projective transformation on the set V such that all vectors in $\mathbb{R}^N(V)$ collinear to the stationary distribution $\pi > 0$ are projected onto a common image point.

Geometric objects, such as points, lines, or planes, can be given a representation as elements in projective spaces based on homogeneous coordinates (Moebius 1827). Any vector of the Euclidean space $\mathbb{R}^N$ can be expanded into $\mathbf{v} = \sum_{k=1}^N \langle \mathbf{v} | \psi_k \rangle \langle \psi_k |$, as well as into the basis vectors

$$\psi_s' \equiv \left(1, \frac{\psi_{s,2}}{\psi_{s,1}}, \ldots, \frac{\psi_{s,N}}{\psi_{s,1}} \right), \quad s = 2, \ldots, N, \tag{2.65}$$

which span the projective space $P\mathbb{R}_\pi^{(N-1)}$,

$$\mathbf{v}\pi^{-1/2} = \sum_{k=2}^N \langle \mathbf{v} | \psi_k' \rangle \langle \psi_k' |,$$

since we always have $\psi_{1,x} \equiv \sqrt{\pi_x} > 0$ for any $x \in V$. The set of all isolated vertices p of the graph $G(V,E)$ for which $\pi_p = 0$ play the role of the plane at infinity, away from which we can use the basis Ψ' as an ordinary Cartesian system. The transition to the homogeneous coordinates (2.65) transforms vectors of $\mathbb{R}^N$ into vectors on the $(N-1)$-dimensional hypersurface $\{\psi_{1,x} = \sqrt{\pi_x}\}$, the orthogonal complement to the vector of stationary distribution π.

2.5.2 Reduction to Euclidean Metric Geometry

The key observation is that in homogeneous coordinates the operator $\widehat{T}^k\big|_{PR_\pi^{(N-1)}}$ defined on the $(N-1)$-dimensional hypersurface $\{\psi_{1,x} = \sqrt{\pi_x}\}$ determines a contractive discrete-time affine dynamical system. The origin is the only fixed point of the map $\widehat{T}^k\big|_{PR_\pi^{(N-1)}}$,

$$\lim_{n \to \infty} \widehat{T}^n \xi = (1, 0, \ldots 0), \tag{2.66}$$

for any $\xi \in PR_\pi^{(N-1)}$, and the solutions are the linear system of points $\widehat{T}^n \xi$ that hop in the phase space (see Fig. 2.4) along the curves formed by collections of points that map into themselves under the repeated action of $\widehat{T}$.

The problem of random walks (2.27, 2.63) defined on finite undirected graphs can be related to a diffusion process which describes the dynamics of a large number of random walkers. The symmetric diffusion process corresponding to the self-adjoint transition operator $\widehat{T}$ describes the time evolution of the normalized expected number of random walkers, $\mathbf{n}(t)\,\pi^{-1/2} \in V \times \mathbb{N}$,

$$\dot{\mathbf{n}} = \widehat{L}\mathbf{n}, \quad \widehat{L} \equiv 1 - \widehat{T} \tag{2.67}$$

where $\widehat{L}$ is the normalized Laplace operator. Eigenvalues of $\widehat{L}$ are simply related to that of $\widehat{T}$, $\lambda_k = 1 - \mu_k$, $k = 1, \ldots, N$, and the eigenvectors of both operators are identical. The analysis of spectral properties of the operator (2.67) is widely used in the spectral graph theory (Chung 1997).

It is important to note that the normalized Laplace operator (2.67) defined on $PR_\pi^{(N-1)}$ is invertible,

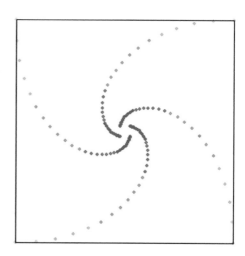

Fig. 2.4 Any vector $\xi \in PR_\pi^{(N-1)}$ asymptotically approaches the origin π under the consecutive actions of the operator $\widehat{T}$

$$\widehat{L}^{-1} = \left(1 - \widehat{T}\right)^{-1}$$
$$= \Sigma_{n \geq 1} \widehat{T}^n.$$
(2.68)

since $\widehat{T}^k\big|_{PR_\pi^{(N-1)}}$ is a contraction mapping for any $k \geq 1$. The unique inverse operator,

$$\widehat{L}^{-1} = \sum_{s=2}^{N} \frac{|\psi'_s\rangle \langle \psi'_s|}{1 - \mu_s},$$
(2.69)

is the Green function (or the Fredholm kernel, the potential of the associated Markov chain) describing long-range interactions between eigenmodes of the diffusion process induced by the graph structure. The convolution with the Green's function gives solutions of inhomogeneous Laplace equations.

In order to obtain a metric on the graph $G(V,E)$, one needs to introduce the distances between points (nodes of the graph) and the angles between lines or vectors by determining the inner product between any two vectors ξ and ζ in $PR_\pi^{(N-1)}$ as

$$(\xi,\zeta)_T = \left(\xi, \widehat{L}^{-1}\zeta\right).$$
(2.70)

The dot product (2.70) is a symmetric real valued scalar function that allows us to define the (squared) norm of a vector $\xi \in PR_\pi^{(N-1)}$ with respect to (w,β) by

$$\|\xi\|_T^2 = \left(\xi, \widehat{L}^{-1}\xi\right).$$
(2.71)

The (nonobtuse) angle $\theta \in [0, 180°]$ between two vectors is then given by

$$\theta = \arccos\left(\frac{(\xi,\zeta)_T}{\|\xi\|_T \|\zeta\|_T}\right).$$
(2.72)

The Euclidean distance between two vectors in $PR_\pi^{(N-1)}$ with respect to (w,β) is defined by

$$\|\xi - \zeta\|_T^2 = \|\xi\|_T^2 + \|\zeta\|_T^2 - 2(\xi,\zeta)_T$$
$$= \mathbb{P}_\xi(\xi - \zeta) + \mathbb{P}_\zeta(\xi - \zeta)$$
(2.73)

where $\mathbb{P}_\xi(\xi - \zeta) \equiv \|\xi\|_T^2 - (\xi,\zeta)_T$ and $\mathbb{P}_\zeta(\xi - \zeta) \equiv \|\zeta\|_T^2 - (\xi,\zeta)_T$ are the lengths of the projections of $(\xi - \zeta)$ onto the unit vectors in the directions of ξ and ζ, respectively. It is clear that $\mathbb{P}_\zeta(\xi - \zeta) = \mathbb{P}_\xi(\xi - \zeta) = 0$ if $\xi = \zeta$.

The spectral representations of the Euclidean structure (2.70–2.73) defined for the graph nodes can be easily derived by taking into account that $\langle i | \psi'_s \rangle = \psi'_{s,i}$, $s = 1, \ldots, N$.

It is obvious from the above formulas that the most important contribution to Euclidean distances defined by (2.73) and (2.71) comes from the second eigenvalue $\mu_2 < 1$. The difference $1 - \mu_2$ is called the spectral gap and defines the bisection of the graph (Cheeger 1969).

2.5.3 Expected Numbers of Steps are Euclidean Distances

The structure of Euclidean space introduced in the previous section can be related to a length structure $V \times V \to \mathbb{R}_+$ defined on a class of all admissible paths P between pairs of nodes in G. It is clear that every path $P(i, j) \in P$ is characterized by some probability to be followed by a random walker depending on the weights $w_{ij} > 0$ of all edges necessary to connect i to j. Therefore, the path length statistics is a natural candidate for the length structure on G.

Let us consider the vector $\mathbf{e}_i = \{0, \ldots 1_i, \ldots 0\}$ that represents the node $i \in V$ in the canonical basis as a density function. In accordance with (2.71), the vector $\mathbf{e}_i$ has the squared norm of $\mathbf{e}_i$ associated to random walks is

$$\| \mathbf{e}_i \|_T^2 = \frac{1}{\pi_i} \sum_{s=2}^{N} \frac{\psi_{s,i}^2}{1 - \mu_s}. \tag{2.74}$$

It is remarkable that in the theory of random walks (Lovasz 1993) the r.h.s. of (2.74) is known as the spectral representation of the first passage time to the node $i \in V$, the expected number of steps required to reach the node $i \in V$ for the first time starting from a node randomly chosen among all nodes of the graph according to the stationary distribution π. The first passage time, $\| \mathbf{e}_i \|_T^2$, can be directly used in order to characterize the level of accessibility of the node i.

The Euclidean distance between any two nodes of the graph G calculated in the $(N-1)$−dimensional Euclidean space associated to random walks,

$$K_{i,j} = \| \mathbf{e}_i - \mathbf{e}_j \|_T^2 = \sum_{s=2}^{N} \frac{1}{1 - \mu_s} \left(\frac{\psi_{s,i}}{\sqrt{\pi_i}} - \frac{\psi_{s,j}}{\sqrt{\pi_j}} \right)^2, \tag{2.75}$$

also gets a clear probabilistic interpretation as the spectral representation of the commute time, the expected number of steps required for a random walker starting at $i \in V$ to visit $j \in V$ and then to return back to i (Lovasz 1993).

The commute time can be represented as a sum, $K_{i,j} = H_{i,j} + H_{j,i}$, in which

$$H_{i,j} = \| \mathbf{e}_i \|_T^2 - (\mathbf{e}_i, \mathbf{e}_j)_T \tag{2.76}$$

is the first-hitting time which quantifies the expected number of steps a random walker starting from the node i needs to reach j for the first time [Lovasz 1993].

The first hitting time satisfies the equation

$$H_{i,j} = 1 + \sum_{i \sim v} H_{v,j} T_{vi} \tag{2.77}$$

reflecting the fact that the first step takes a random walker to a neighbor $v \in V$ of the starting node $i \in V$, and then it must reach the node j from there. In principle, the latter equation can be directly used for the computation of the first hitting times; however, $H_{i,j}$ is not the unique solution of (2.77); the correct definition requires an appropriate diagonal boundary condition, $H_{i,i} = 0$, for all $i \in V$ (Lovasz 1993). The spectral representation of $H_{i,j}$ given by

$$H_{i,j} = \sum_{s=2}^{N} \frac{1}{1-\mu_s} \left(\frac{\psi_{s,i}^2}{\pi_i} - \frac{\psi_{s,i}\psi_{s,j}}{\sqrt{\pi_i \pi_j}} \right),$$ (2.78)

seems much easier to calculate. From the obvious inequality $\lambda_2 \leq \lambda_r$, it follows that the first-passage times are asymptotically bounded by the spectral gap, namely $\lambda_2 = 1 - \mu_2$.

The matrix of first hitting times is not symmetric, $H_{ij} \neq H_{ji}$, even for a regular graph. However, a deeper triangle symmetry property (see Fig. 2.5) has been observed by Coppersmith et al. (1993) for random walks defined by the transition operator (2.27). Namely, for every three nodes in the graph, the consequent sums of the first hitting times in the clockwise and the counterclockwise directions are equal,

$$H_{i,j} + H_{j,k} + H_{k,i} = H_{i,k} + H_{k,j} + H_{j,i}.$$ (2.79)

We can now use the first hitting times in order to quantify the accessibility of nodes and subgraphs for random walkers.

It is clear from the spectral representations given above that the average of the first hitting times with respect to its first index is nothing else, but the first passage time to the node,

$$\| e_i \|_T^2 = \sum_{j \in V} \pi_j H_{j,i}.$$ (2.80)

The average of the first hitting times with respect to the second index is called the random target access time (Lovasz 1993). It quantifies the expected number of steps required for a random walker to reach a randomly chosen node in the graph (a target). In contrast to (2.80), the random target access time $\mathfrak{I}_G$ is independent of the starting node $i \in V$ being a *global* spectral characteristic of the graph,

$$\begin{aligned} \mathfrak{I}_G &= \sum_{j \in V} \pi_j H_{i,j} \\ &= \sum_{k=2}^{N} \frac{1}{1-\mu_k}. \end{aligned}$$ (2.81)

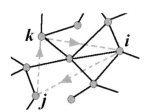

Fig. 2.5 The triangle symmetry of the first hitting times: the sum of first hitting times calculated for random walks defined by (2.27) visiting any three nodes i, j, and k, equals the sum of the first hitting times in the reversing direction

The latter equation expresses the so-called random target identity (Lovasz 1993).

Finally, the scalar product $(\mathbf{e}_i, \mathbf{e}_j)_T$ estimates the expected overlap of random paths toward the destination nodes i and j starting from a node randomly chosen in accordance with the stationary distribution of random walks π. The normalized expected overlap of random paths given by the cosine of an angle calculated in the $(N-1)$–dimensional Euclidean space associated to random walks has the structure of Pearson's coefficient of linear correlations that reveals its natural statistical interpretation. If the cosine of the angle (2.72) is close to 1 (zero angles), we conclude that the expected random paths toward both nodes are mostly identical. A value of cosine close to -1 indicates that the walkers share almost the same random paths, but in opposite directions. The correlation coefficient is near 0 if the expected random paths toward the nodes have a very small overlap. As usual, the correlation between nodes does not necessarily imply a direct causal relationship (an immediate connection) between them.

2.5.4 Probabilistic Topological Space

In the previous sections we have shown that, given a symmetric affinity function $w : V \times V \to \mathbb{R}_+$, we can always define an Euclidean metric on V based on the first-access time properties of standard stochastic process, the random walks defined on the set V with respect to the matrix $w_{ij} \geq 0$.

In particular, we can introduce this metric on any undirected graph $G(V,E)$ converting it in a metric space. The Euclidean distance interpreted as the commute time induces the metric topology on $G(V,E)$. Namely, we define the open metric ball of radius r about any point $i \in V$ as the set

$$B_r(i) = \left\{ j \in V : K_{i,j}^{1/2} < r \right\}. \tag{2.82}$$

These open balls generate a topology on V, making it a topological space. A set U in the metric space is open if and only if for every point $i \in U$ there exists $\varepsilon > 0$ such that $B_\varepsilon(r) \subset U$, (Burago et al. 2001). Explicitly, a subset of V is called open if it is a union of (finitely or infinitely many) open balls.

2.5.5 Euclidean Embedding of the Petersen Graph

We consider the Euclidean embedding of the Petersen graph (see Fig. 2.6) by random walks as an example.

The Petersen graph is a regular graph, $k_i = 3$, $i = 1, \ldots 10$, consisting of 10 nodes and 15 edges, $\sum_i k_i = 30$. It constitutes a notorious example for the theory of complex network since the graph nodes cannot be distinguish by its standard methods.

The stationary distribution of random walks on the graph nodes is uniform, $\pi_i^{(\text{Pet})} = 0.1$. The spectrum of the random walk transition operator (2.44) defined

Fig. 2.6 The Petersen graph

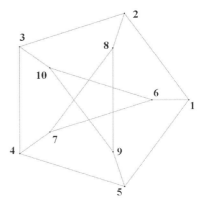

on the Petersen graph consists of the Perron eigenvalue, $\mu_1 = 1$ which is simple, then the eigenvalue $\mu_2 = 1/3$ with multiplicity 5, and $\mu_3 = -2/3$ with multiplicity 4. Therefore, in the linear vector space of eigenvectors, there are just three linearly independent eigenvectors, and two eigensubspaces for which the orthonormal basis vectors can be calculated, so that the matrix of basis vectors which we use in (2.74 – 2.75) always has full column dimension.

Random walks embed the Petersen graph into nine-dimensional Euclidean space, in which all nodes have equal norm (2.74), $\|i\|_T = 3.1464$ meaning that the expected number of steps required to reach a node equals 9.9.

Indeed, the structure of nine-dimensional vector space induced by random walks defined on the Petersen graph cannot be represented visually, however if we choose one node as a point of reference, we can draw its two-dimensional projection by arranging other nodes at distances calculated according to (2.75), and under the angles (2.72) they are with respect to the chosen reference node (see Fig. 2.7).

It is expected that, on average, a random walker starting at node #1 visits any peripheral node (#2, 3, 4, 5) and then returns back in 18 random steps, while 24 random steps are expected for visiting any node in the central component of the graph (#6, 7, 8, 9, 10). Due to the symmetry of the Petersen graph, the diagram displayed

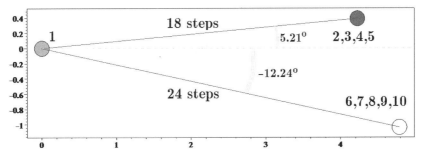

Fig. 2.7 The two-dimensional projection of the Euclidean space embedding of the Petersen graph drawn with respect to the node #1

in Fig. 2.7 would be essentially the same if we draw it with respect to any other peripheral node (#2, 3, 4, 5). However, it appears to be mirror-reflected if we draw the figure taking any internal node (#6, 7, 8, 9, 10) as the origin. Therefore, we can conclude that the Petersen graph contains two components – the periphery and the core – whose nodes appear to be as much as one-quarter more isolated (18 random steps vs. 24 random steps) than those belonging to the same group.

The positive and negative angles between the nodes belonging to the different components of the Petersen graph indicate that paths of random walkers travelling toward destination nodes essentially, have the same component overlap while they are loosely overlapped and mostly run in opposite directions if they follow the alternative components.

In Fig. 2.8, we have presented the Euclidean space embedding of the Petersen graph by means of the scalar product matrix, $S_{ij} = (\mathbf{e}_i, \mathbf{e}_j)_T$. The diagonal elements of S_{ij} are the first-passage times to the corresponding nodes in the graph, while the entries out of the diagonal give the expected overlaps of random paths toward i and j. The eigenvectors belonging to the largest eigenvalues of the matrix S_{ij} delineate those directions in the vector space along which the scalar product $(\mathbf{e}_i, \mathbf{e}_j)_T$ has the largest variance. We can represent each node of the Petersen graph by a point in three-dimensional space by regarding the corresponding components of three major eigenvectors as its Cartesian coordinates. It is then clear from Fig. 2.8 that the Petersen graph is bisected in the probabilistic Euclidean space associated with random walks.

In the next section, we apply the method of structural analysis for exploring city spatial graphs.

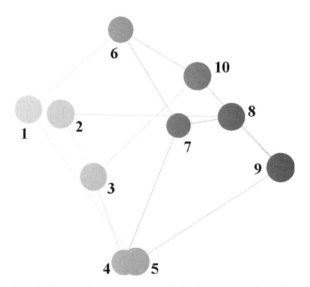

Fig. 2.8 The 3-D representation of the Petersen graph in the Euclidean space associated with random walks

2.6 Case study: Affine Representations of Urban Space

Any connected undirected graph, no matter how complex its structure is, constitutes a normed metric probability space in which the first-passage time to the nodes
defines the system of length (Blanchard et al. 2008).

The traffic flow forecasting of first-passage times have been recently studied
(Sun et al. 2005) in order to model the wireless terminal movements in a cellular
wireless network (Jabbari et al. 1999), in a statistical test for the presence of a random walk component in the repeat sales price index models in house prices [Hill
et al. 1999], in the growth modelling of urban agglomerations [Pica et al. 2006], and
in many other works where the impact of random walks on city plans and physical
landscapes has been considered.

In contrast to all previous studies, in our book, we use discrete time random walks
in order to investigate the configuration of urban places represented by means of the
spatial graph. In particular, in the present section, we use the first-passage times in
order to estimate the levels of accessibility of compact urban patterns. The physical
distances between certain locations are of no matter in such a representation and,
therefore, random walks are the natural tool for the investigation since, at each time
step, a random walker moves to a neighboring place independently of how far it is.

2.6.1 Ghetto of Venice

The spatial network of Venice that stretches across 122 small islands is comprised
of 96 canals which serve as roads.

In March 1516 the Government of the Serenissima Repubblica issued special
laws, and the first Ghetto of Europe was instituted in the Cannaregio district, the
northernmost part of the city. It was the area where Jews were forced to live and
not leave from sunset to dawn. Surrounded by canals, this area was only linked to
the rest of the city by two bridges. The quarter had been enlarged later to cover
the neighboring Ghetto Vecchio and the Ghetto Nuovissimo. As a result a specific
Gehtto canal sub-network arose in Venice weakly connected to the main canals.

The Ghetto existed for more than two and one-half centuries, until Napoleon
conquered Venice and finally opened and eliminated every gate (1797). Despite the
fact that the political and religious grounds for the ghettoization of these city quarters have disappeared, these components are still relatively isolated from the major city canal network that can be spotted by estimating the first-passage and first
hitting times in the network of Venetian canals. Computations of the first hitting
times between street and canals in the compact urban patterns have been reported in
Volchenkov et al. (2007a).

In Fig. 2.9, we have shown the matrix plot of the first hitting times to the nodes
of the spatial graph for 96 canals in the city canal network of Venice.

The variance of the first hitting times to the nodes could help us estimate the
quality of spatial representations of urban networks that we use. Already the visual
analysis of the variances of the first hitting times to the Venetian canals shows (see

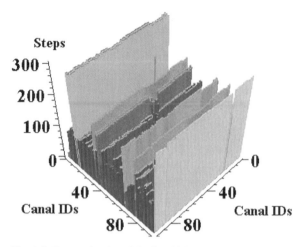

Fig. 2.9 The matrix plot of the first hitting times between nodes of the spatial graph for 96 canals in the city canal network of Venice

Fig. 2.9) that the values $H_{i,j}$ vary slightly over the second index j and, therefore, the averaged first-hitting time (i.e., the first-passage time) to a canal,

$$\|\mathbf{e}_i\|_T^2 = \sum_{j=1}^{N} \pi_i H_{i,j}, \tag{2.83}$$

can be used as a measure of its accessibility from other canals in the canal network.

The probability distribution of the first-passage times,

$$P(x) = \Pr \left[i \in G | \|\mathbf{e}_i\|_T^2 = x \right], \tag{2.84}$$

allows us to explore the connectedness of the entire canal network in the city. In particular, if the graph contains either groups of relatively isolated nodes, or bottle-necks, they can be visually detected on the probability distribution profile (2.84). This method can facilitate the detection of urban ghettos and sprawl.

The distribution of canals over the range of the first-passage time values is represented by a histogram shown in Fig. 2.10). The height of each bar in the histogram represents the number of canals in the canal network of Venice for which the first-passage times fall into the disjoint intervals (known as bins).

It is fascinating that while most Venetian canals can probably be reached from everywhere in 300 random steps, almost 600 random steps are required in order to reach those canals surrounding the Venetian Ghetto.

It is important to stress the essential difference between the first-passage time to a node as a measure quantifying the global property of the node with respect to other nodes in the graph and the classical integration measure (1.12) related to the simple mean distance ℓ_i from the node i to any other node in the graph used in the traditional space syntax approach.

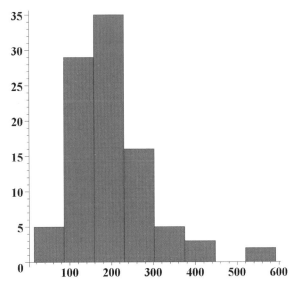

Fig. 2.10 The histogram of the distribution of canals over the range of the first-passage times in the Venetian canal network

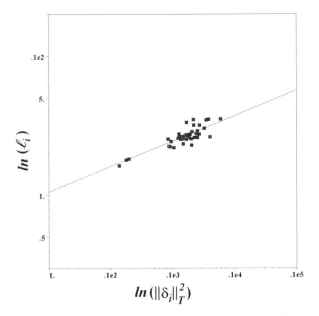

Fig. 2.11 The scatter plot (in the log-log scale) of the mean distances vs. the value of first-passage times for the network of Venetian canals. The plot indicates a slight, but positive relation between these two characteristics: the slope of regression line equals 0.18. Three data points characterized by the shortest first-passage times and the shortest mean distances represent the main water routes of Venice: the Lagoon, the Giudecca canal, and the Grand canal

The relation between the mean distance (1.8) and the first-passage time to it (2.74) is very complicated and strongly depends on the topology of the graph. In Fig. 2.11, we have shown the scatter plot (in the log-log scale) of the mean distances vs. the value of first-passage times for the network of Venetian canals.

2.6.2 Spotting Functional Spaces in the City

Random walks defined on a connected undirected graph partition its nodes into equivalence classes according to their accessibility for random walkers.

The triangle symmetry property (2.79) can be used in order to classify nodes with regard to their accessibility levels (Lovasz 1993). Namely, nodes can be ordered so that $i \in V$ precedes $j \in V$ if and only if $H_{i,j} \leq H_{j,i}$. Such a relative ordering can be obtained by fixing any $i \in V$ as a reference node of the graph and then by estimating all other nodes according to the first hitting times difference value,

$$
\begin{aligned}
\eth_{ij} &= H_{j,i} - H_{i,j} \\
&= \left\| \mathbf{e}_j \right\|_T^2 - \left\| \mathbf{e}_i \right\|_T^2 .
\end{aligned}
\tag{2.85}
$$

Indeed such an ordering is by no means unique, because of the ties. However, if we partition all nodes in the graph by putting $i \in V$ and $j \in V$ into the same equivalence class when their reciprocal first hitting times are equal, $\eth_{ij} \simeq 0$, (this is an equivalence relation, see Lovasz (1993), then there is a well-defined ordering of equivalence classes, which is obviously independent of any particular reference node $i \in V$. Let us mention that, in general, the accessibility equivalence classes do not form connected subgraphs of the initial graph.

The marginal accessibility classes can be of essential practical interest. The nodes in the best accessibility class (characterized by the minimal first hitting time) are easy to reach, but difficult to leave – they act as traps in the graph. The city locations related to public processes of trade, exchange, and government tend to occupy those places which can be easily reached from everywhere. At the same time, in order to increase the chance of plausible contacts between people, it seems reasonable that people could stay in these places longer as if being trapped there. Then, the city locations belonging to the best accessibility class would naturally provide the best places for the public process.

Alternatively, the nodes from the worst accessibility class (the hidden places, characterized by the largest first hitting times) are difficult to reach, but very easy to get out. If found on the city spatial graph, they constitute the optimal location for a residential area where the occasional appearance of strangers is unwilling.

2.6.3 Bielefeld and the Invisible Wall of Niederwall

The properties of first hitting times can be used in order to estimate the accessibility of certain streets and districts in the city.

The distribution of first hitting times in the downtown of Bielefeld is of interest since it reveals two structurally different parts (see Fig. 2.12, left) – part "A" keeps its original structure, while part "B" has been subjected to partial redevelopment. Niederwall, the central itinerary crossing the downtown of Bielefeld, constitutes a natural boundary between two parts of the city and conjugates them both.

Computations of first hitting times to the streets in the studied compact urban structures abstracted as spatial graphs in the framework of the street-named approach convinced us that for any given node i the first hitting times to it, H_{ij}, fluctuate slightly with respect to the second (destination) index j in comparison with their typical values and, therefore, the simple partial mean first hitting times,

$$
\begin{aligned}
h_i(A \to A) &= N_A^{-1} \sum_{j \in A} H_{ij}, \quad i \in A, \\
h_i(A \to B) &= N_B^{-1} \sum_{j \in B} H_{ij}, \quad i \in A, \\
h_i(B \to B) &= N_B^{-1} \sum_{j \in B} H_{ij}, \quad i \in B, \\
h_i(B \to A) &= N_A^{-1} \sum_{j \in A} H_{ij}, \quad i \in B,
\end{aligned}
\tag{2.86}
$$

in which N_A and N_B are the total number of locations in the A and B parts of the urban pattern in the downtown of Bielefeld consequently, and can be considered good empirical parameters estimating the mutual accessibility of a location within the different parts of the city (Volchenkov et al. 2007a).

Then, the distributions of the simple partial mean first hitting times in the city,

$$
\alpha_h(A/B) = \Pr[h_i(A/B \to A/B) = h],
\tag{2.87}
$$

can be considered as the empirical estimation of its connectedness.

In Fig. 2.12, we have displayed the distributions of the simple partial mean first hitting times (2.86) to the streets located in the medieval part "A" starting from those located in the same part of Bielefeld downtown, from "A" to "A" (solid line). This has been computed by averaging H_{ij} over $i, j \in A$ in (2.86). The dashed line represents the distribution of mean access times to the streets located in the modernized part "B" starting from the medieval part "A" (from "A" to "B", $i \in A$ and $j \in B$). One can see that, on average, it takes longer to reach the streets located in "B" starting from "A." Similar behavior is demonstrated by the random walkers starting from "B" (see Fig. 2.13): on-average, it requires more time to leave a district for another one. Study of random walks defined on the dual graphs helps to detect the quasi-isolated districts of the city.

A sociological survey shows that up to 85,000 people arrive in the city of Bielefeld at the central railway station next to the ancient part "A" during weekends. While exploring the downtown of the city, they eventually reach Niederwall and then usually return to part "A" as if the structural dissimilarity between two parts of the city center clearly visible along Niederwall was an invisible wall. The total number of travellers in part "B" during weekends usually does not exceed 35,000 people, although there are many more old buildings that are potentially attractive to tourists preserved in part "B" than in "A."

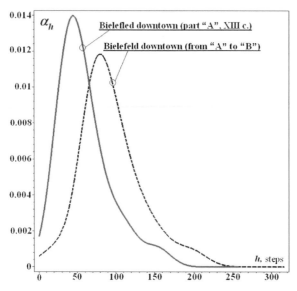

Fig. 2.12 The distributions, of the simple partial mean first hitting times to the streets located in the medieval part "A" starting from those located in the same part of Bielefeld downtown, from "A" to "A" (solid line). The dashed line presents the distribution of mean access times to the street located in the modernized part "B" starting from the medieval part "A" (from "A" to "B"). On average, in takes longer to reach the streets located in "B" starting from "A"

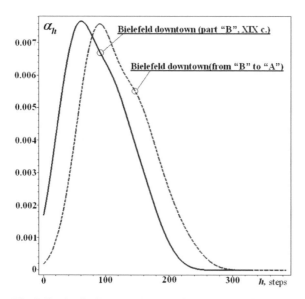

Fig. 2.13 The distributions of the simple partial mean first hitting times to the streets located in the "B" part starting from "B" (*solid line*). The dashed line represents the distribution of mean access times to the street located in the "A" part starting from "B"

2.6.4 Access to a Target Node and the Random Target Access Time

The notion of isolation acquires the statistical interpretation by means of random walks. The first-passage times in the city vary strongly from location to location. Those places characterized by the shortest first-passage times are easy to reach while many random steps would be required in order to get into a statistically isolated site.

Being a global characteristic of a node in the graph, the first-passage time assigns absolute scores to all nodes based on the probability of paths they provide for random walkers. The first-passage time can, therefore, be considered a natural statistical centrality measure of the vertex within the graph.

The possible relation between the local and global properties of nodes is the most profound feature of a complex network. It is intuitive that the the first-passage time to a node, $\|\mathbf{e}_i\|_T^2$, has to be positively related to the time of recurrence r_i: the faster a random walker hits the node for the first time, the more often he is expected to visit it in the future. This intuition is supported by the expression (2.74) from which it follows that $\|\mathbf{e}_i\|_T^2 \propto r_i$ provided the sum

$$\sum_{s=2}^{N} \frac{\psi_{s,i}^2}{(1 - \mu_s)} \simeq \text{Const} \tag{2.88}$$

uniformly for all nodes. The relation (2.88) is by no means a trivial mathematical fact, as we shall see below.

We consider again two different types of random walks personified by Alice and Bob in Sect. 2.4. Let us recall that Alice performs the unbiased random walk defined by the transition operator (2.51) implying no preference between nodes. In contrast to her, while executing his biased random walks defined by (2.52), Bob follows "a strategy" paying attention primarily to those nodes of the highest betweenness centrality.

In Fig. 2.14, we have represented the two-dimensional projection of the probabilistic Euclidean space of 355 locations in Manhattan (New York) set up by the unbiased random walk performed by Alice. Nodes of the spatial graph are shown by disks with radiuses $\rho \propto k_i$ taken proportional to connectivity of the places. Broadway, a wide avenue in Manhattan which also runs into the Bronx and Westchester County, possesses the highest connectivity and, therefore, is located at the center of the diagram shown in Fig. 2.14. Other places have been located at their Euclidean distances (i.e., the first hitting times) from Broadway calculated accordingly

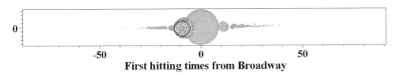

First hitting times from Broadway

Fig. 2.14 The two-dimensional projection of the probabilistic Euclidean space of 355 locations in Manhattan (New York) from Broadway set up by the unbiased random walk performed by Alice

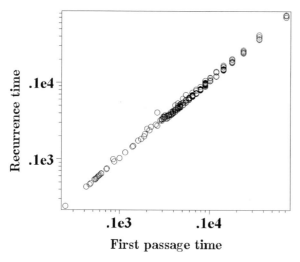

Fig. 2.15 In the unbiased random walk performed by Alice in Manhattan, the times of recurrence to locations scale linearly with the first-passage times to them

to (2.75), and at the angles calculated by (2.72). It is remarkable that the diagram in Fig. 2.14 displays a nice ordering of places in Manhattan with respect to their connectivity k_i: the less connected the place, the longer its first hitting time from Broadway.

In the unbiased random walk performed by Alice in Manhattan, the local property of an open space is qualified by its connectivity k_i which determines the recurrence time of random walks into it (2.55), and the global property of the location is estimated by the first-passage time to it calculated accordingly (2.74). These local and global properties appear to be strongly positively related for every location in the city. In Fig. 2.15, we have presented the log-log plot exhibiting the linear scaling between the first-passage times and recurrence times for the unbiased random walk of Alice.

The relation (2.88) apparently holds for Alice's walks. We can conclude that while the first eigenvector ψ_1 belonging to the largest eigenvalue of the transition operator (2.51) describes the local connectivity of nodes; all other eigenvectors report the global connectedness of the graph.

However, the same correlation is not true for the biased random walks (2.52) performed by Bob.

In Fig. 2.16, we have sketched the central fragment of the two-dimensional projection of Manhattan (New York) set up by the biased random walk of Bob. Radiuses of disks shown in Fig. 2.16 are proportional to the betweenness centrality indices of the correspondent city locations (which are inverse proportional to the recurrence times, in Bob's random walk). The betweenness centrality of Broadway is the highest one, and other places are located at their first hitting times from Broadway calculated according to (2.75) and at the angles calculated by (2.72) for the biased random walks of Bob (2.52). Despite the interesting structural patterns visi-

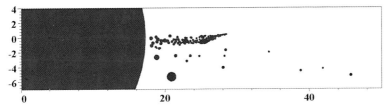

Fig. 2.16 The central fragment of the two-dimensional projection of the probabilistic Euclidean space of 355 locations in Manhattan (New York) from Broadway set up by the biased random walk performed by Bob

ble in Fig. 2.16, it is, nevertheless, clear that in Bob's biased random walk there is no direct positive relation between the recurrence times and the first hitting times reported for the different city locations.

The average of the first hitting times $H_{i,j}$ with respect to its second index expresses the expected number of steps a random walker needs to reach an arbitrary node of the graph (a target) chosen randomly from the stationary distribution of random walks π.

The random target access time can be easily computed accordingly (2.81) provided the spectrum of the random walk transition operator is known. In Fig. 2.17, we have displayed the comparative diagram indicating the values of random target access times for the five compact urban patterns. Beside these bars, we have displayed others with heights equal to the sizes of the studied city spatial graphs. Obviously, the random target access time is an extensive quantity that grows with the size of the city.

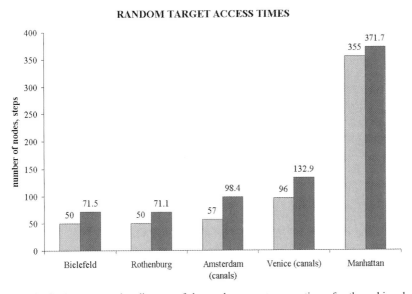

Fig. 2.17 The comparative diagram of the random target access times for the unbiased random walks and the sizes of spatial graphs for five compact urban patterns. Heights of left bars show the number of nodes in the graphs. Bars on the right indicate the random target access times

2.6.5 Pattern of Spatial Isolation in Manhattan

The character and development of Manhattan, the acknowledged heart of New York City, are essentially shaped by geography – Manhattan had only been linked to the other boroughs by bridges and tunnels at the end of the 19th Century. Originally settled around the southern tip dominating New York Harbor, Manhattan expanded northward and encompassed the upper East and West Sides in the mid-19th Century, and the fields above Central Park were settled near the turn of the 19th and 20th Centuries.

The spatial graph representing the structure of the urban pattern in Manhattan constitutes a Euclidean space of dimension 354 over the field of real numbers, endowed with an inner product (2.70). Gram's matrix is a square matrix,

$$\mathscr{G}_{i,j} = (\mathbf{e}_i, \mathbf{e}_j)_T, \tag{2.89}$$

consisting of pairwise scalar products of elements (vectors) representing nodes in this Euclidean space. All Gram matrices are nonnegative definite, and may be positive definite if all vectors in scalar products are linearly independent. The diagonal elements $\|\mathbf{e}_i\|_T^2$ of the Gram matrix (2.89) define the first-passage times to all places in Manhattan, and the off-diagonal elements, $(\mathbf{e}_i, \mathbf{e}_j)_T$, $i \neq j$, are nothing else but the expected overlaps between all pairs of random paths leading to i and j.

The Gram matrix (2.89) can be used as a similarity matrix in construction of the three-dimensional visual representations of the spatial graphs of urban area networks. Namely, we can solve the eigenvalue problem,

$$\mathscr{G}\mathbf{u} = \alpha\mathbf{u}, \tag{2.90}$$

and use the components of the first triple of eigenvectors $\mathbf{u}_1$, $\mathbf{u}_2$, and $\mathbf{u}_3$ belonging to the three largest eigenvalues $\alpha_1 > \alpha_2 > \alpha_3$ of the matrix $\mathscr{G}$. These eigenvectors determine the directions in which the spatial graph of the urban area network exhibits the maximal structural similarity. Then, we can use the components of these eigenvectors $(u_{1,i}, u_{2,i}, u_{3,i})$ to define the coordinates of points representing the spatial locations (nodes) in a visual three-dimensional representation of the spatial graph.

The three-dimensional image of the spatial graph comprising 355 locations in the urban pattern of Manhattan is presented in Fig. 2.18. The radiuses of balls representing the individual places have been taken as proportional to their degrees (the numbers of other locations they are adjacent to in the urban pattern of Manhattan).

The visualization of the affine representation for the urban area network in Manhattan reveals its neighborhood structure, since if random paths originated from a randomly chosen node of the spatial graph and destined towards two different locations are expected to be mostly overlapped, the vertices corresponding to them also appear to be close in the graph representation shown in Fig. 2.18. The groups of places allocated closely in the graph of Fig. 2.18 may represent the geographically localized structural communities with highly probable face-to-face interactions among their dwellers, i.e., neighborhoods.

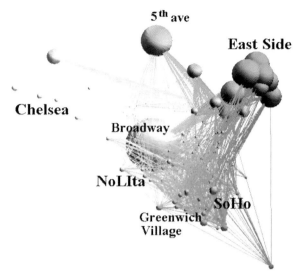

Fig. 2.18 The three-dimensional representation of the spatial graph of Manhattan based on the first three eigenvectors of the Gram matrix (2.89) belonging to its three maximal eigenvalues. The radiuses of balls representing the individual spatial locations are taken proportional to their degrees

However, the spatial graph of Manhattan (Fig. 2.18) is rather complicated to be analyzed visually. In order to understand its structure, we use the method of kernel density estimations (Wasserman 2005) for the data of first-passage times to all graph nodes. The curve in Fig. 2.19 is the probability density function of the first-passage times (2.84) calculated over the urban pattern of Manhattan.

The first-passage times probability density curve enables us to classify all places in the spatial graph of Manhattan into four groups according to their first-passage

The first group of locations is characterized by the minimal first-passage times; they are probably reached for the first time from any other place of the urban pattern in just 10–100 random navigational steps. These locations are identified as belonging to the downtown Manhattan (at the south and southwest tips of the island) – the Financial District and Midtown Manhattan. These neighborhoods are roughly coterminous with the boundaries of the ancient New Amsterdam settlement founded in the late 17th Century. Both districts comprise the offices and headquarters of many of the city's major financial institutions such as the New York Stock Exchange and the American Stock Exchange (in the Financial District). Federal Hall National Memorial, which had been anchored by the World Trade Center until the September 11, 2001 terrorist attacks is also encompassed in this area. We might conclude that the group of locations characterized by the best structural accessibility is the heart of the public process in the city (Hillier et al. 1984).

The neighborhoods from the second group comprise the locations that can be reached for the first time in several hundreds to roughly 1,000 random navigational steps from any other place of in urban pattern. In Fig. 2.19, we have marked this zone as a city core. SoHo (to the south of Houston Street), Greenwich Village,

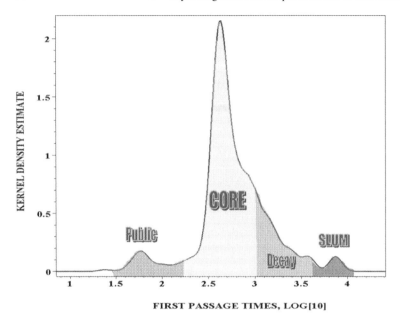

Fig. 2.19 The probability density function of first-passage times in Manhattan.

Chelsea (Hell's Kitchen), the Lower East Side, and the East Village are among them – they are commercial in nature and known for upscale shopping and the "Bohemian" lifestyle of their dwellers which contributes into New York's art industry and nightlife.

The relatively isolated neighborhoods, such as the Bowery, some segments in Hamilton Heights and Hudson Heights, Manhattanville (bordered on the south by Morningside Heights), TriBeCa (Triangle Below Canal) and some others can be associated to the third structural category as being reached for the first time from 1,000 to 3,000 random steps starting from a randomly chosen place in the spatial graph of Manhattan. In Fig. 2.19, this category belongs to the decaying tail of the distribution curve.

Interestingly, many locations belonging to the third structural group comprise the diverse and eclectic mix of different social and religious groups. Many famous houses of worship were established there during the late 19th Century – St. Mary's Protestant Episcopal Church, Church of the Annunciation, St. Joseph's Roman Catholic Church, and Old Broadway Synagogue in Manhattanville are among them.

The neighborhood of the Bowery in the southern portion of Manhattan had been most often associated with the poor and the homeless. From the early 20th Century, the Bowery became the center of the so-called "b'hoy" subculture of working-class young men frequenting the cruder nightlife. Petty crime and prostitution followed in their wake, and most respectable businesses, the middle-class, and entertainment had fled the area. Today, the dramatic decline has lowered crime rates in the district to a level not seen since the early 1960s and it continues to fall. Although a zero-tolerance policy targeting petty criminals is being heralded as a major reason for the decrease in crime, no clear explanation for the crime rate's fall has been found.

The last structural category comprises the most isolated segments in the city, mainly allocated in Spanish and East Harlem. They are characterized by the longest first-passage times from 5,000 to more than 7,000 of random steps on the spatial graph in Fig. 2.18. Structural isolation is fostered by the unfavorable confluence of many factors such as the close proximity to Central Park (an area of 340 hectares removed from the otherwise regular street grid), the boundness by the strait of Harlem River separating the Harlem and the Bronx, and the remoteness from the main bridges (the Triborough Bridge, the Willis Avenue Bridge, and the Queensboro Bridge) that connect the boroughs of Manhattan to the urban arrays in Long Island City and Astoria.

Many social problems associated with poverty, from crime to drug addiction, have plagued the area for some time. The haphazard change of the racial composition of the neighborhood occurred at the beginning of the 20th Century together with the lack of adequate urban infrastructure and services fomenting racial violence in deprived communities and made the neighborhood unsafe – Harlem became a slum. The neighborhood had suffered with unemployment, poverty, and crime for quite a long time and even now, despite the sweeping economic prosperity and redevelopment of many sections in the district, the core of Harlem remains poor.

A six-color visual pattern displayed in Fig. 2.20 represents the pattern of structural isolation (quantified by the first-passage times) in Manhattan (darker color corresponds to longer first-passage times). The spatial distribution of isolation in the urban pattern of Manhattan (Fig. 2.20) shows a qualitative agreement with the data on the crime rates in the borough collected in the framework of the Uniform Crime Reporting (UCR) program by the Disaster Center in association with the Rothstein Catalog on Disaster Recovery (USA). Although the UCR data do not include a record of every crime reported to law enforcement, the most accurate number of crimes reported are those involving death, and the least accurate is the number of rapes that are reported. Usually, the high rates of crime are reported from the locations in areas with a large industrial zone or those serving as tourist destinations.

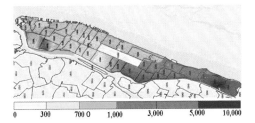

| 0 | 300 | 700 0 | 1,000 | 3,000 | 5,000 | 10,000 |

Fig. 2.20 Isolation map of Manhattan. Isolation is measured by first-passage times to the places. Darker color corresponds to longer first-passage times

According to the latest U.S. Census data (U.S.Census 2006), Manhattan residents have the highest average income ($73,000) in the United States. Indeed, this average is seriously skewed because of the highest housing costs and other cost of living expenditures that keep pace with the wave of real estate development in a city experiencing tremendous growth. The life and work in many Manhattan neighborhoods, as well as their shapes, are undergoing rapid transformation, although incomes of the majority of New Yorkers are not keeping up with the cost of living.

The mean household income in Manhattan demonstrates a striking spatial pattern which can be analyzed and compared with the pattern of structural isolation quantified by the first-passage times (see Fig. 2.21). We have used the open data reported by the Pratt Center for Community Development (NY) on the mean household income in Manhattan per year for 2003 (indicated by bars in Fig. 2.21) and identified their locations on the isolation map of the city. Although each income bar in general covers a wide range of isolation values, and, therefore, the juxtaposing bars are partially overlapped, the graph in Fig. 2.21 displays a clear tendency positively relating income and the level of accessibility to a neighborhood.

The Justice Mapping Center (JMC) (Brooklyn, NY) uses computer mapping and other graphical depictions of quantitative data to analyze and communicate social policy information. Their approach is based on the plain fact that American society is stratified and that justice, social welfare, and economic development policies are intimately related to particular jurisdictions and neighborhoods, where people of those different social and economic strata live. The analysis of high resettlement neighborhoods in New York City performed by the JMC in October 2006 convincingly shows that the distributions of crime and prison expenditures in Manhattan

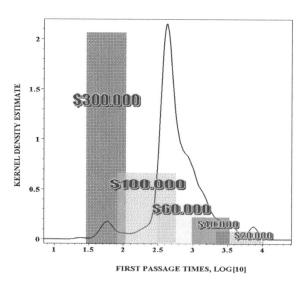

Fig. 2.21 Who makes the most money in Manhattan? (2003)

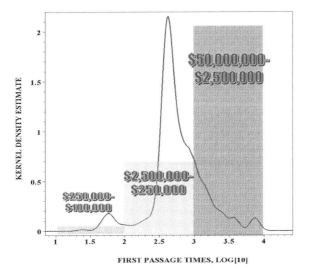

Fig. 2.22 Prison expenditures in Manhattan districts per year (2003)

exhibit a clear spatial pattern. In Fig. 2.22, we have presented the results of comparison between the spatial pattern of prison expenditures in different zip code areas of Manhattan dated to 2003, and the pattern of structural isolations.

The data shown on the diagram Fig. 2.22 positively relates the level of structural isolation of a neighborhood to its prison expenditures.

The relation between the extent of structural isolation and the specified reference levels of prison expenditures can be measured in a logarithmic scale by using decibels (dB). When referring to estimates of isolation by means of first-passage times (FPT), a ratio between two levels inherent to the different locations A and B can be expressed in decibels by evaluating,

$$I_{AB} = 10 \log_{10} \left(\frac{\mathrm{FPT}(A)}{\mathrm{FPT}(B)} \right), \tag{2.91}$$

where FPT(A) and FPT(B) ,are the first-passage times to A and B, respectively. In Fig. 2.22, all bars representing the amount of prison expenditures in neighborhoods of Manhattan are of the same width corresponding to the increase of isolation by $I_{AB} = 10\,\mathrm{dB}$. Similarly, the decibel units can be applied to express the relative growth in prison expenditures (PE) between neighborhoods A and B,

$$PE_{AB} = 10 \log_{10} \left(\frac{\mathrm{PE}(A)}{\mathrm{PE}(B)} \right), \tag{2.92}$$

The data of Fig. 2.22 represented as a table (see Table 2.1) below demonstrate that prison expenditures show a tendency to increase as isolation worsens.

Table 2.1 Growth of prison expenditures as isolation worsens

Location category	Increase of isolation (dB)	Prison expenditure growth (dB)
Financial district	10	3.98
City core	10	10
Decay & Slums	10	13.01

2.6.6 Neubeckum: Mosque and Church in Dialog

Churches are buildings used as religious places, in the Christian tradition. In addition to being a place of worship, the churches in Western Europe were utilized by the community in other ways, i.e., as a meeting place for guilds. Typically, their location was the focus of a neighborhood, or a settlement.

Today, because of the intensive movement of people between countries, the new national unities out of cultural and religious diversity have appeared. United State's rich tradition of immigrants has demonstrated the ability of an increasingly multicultural society to unite different religious, ethnic and linguistic groups into the fabric of the country, and many European countries have followed suit. (Portes et al. 2006).

Religious beliefs and institutions have and continue to play a crucial role in new immigrant communities. Religious congregations often provide ethnic, cultural and linguistic reinforcements, and often help newcomers to integrate by offering a connection to social groups that mediate between the individual and the new society, so that immigrants often become even more religious in their new country of residence (Kimon 2001).

It is not surprising that the buildings belonging to religious congregations of newly arrived immigrants are usually located not at the centers of cities in the host country – the changes in function result in a change of location. In Sect. 2.6.5, we have discussed that religious organizations of immigrants in the urban pattern of Manhattan have been usually founded in the relatively isolated locations, apart from the city core, like those in Manhattanville. We have seen that the typical first-passage times to the "religious" places of immigrant communities in Manhattan scale from 1,000 to 3,000 random steps. It is interesting to also check this observation for the religious congregation buildings of recent immigrants in Western Europe.

Despite the mosque and the church being located in close geographic proximity in the city of Neubeckum (see Fig. 2.23), their locations are dramatically different with respect to the entire city structure. The analysis of the spatial graph of the city of Neubeckum by random walks shows that, while the church is situated in a place belonging to the city core, and just 40 random steps are required in order to reach it for the first time from any arbitrary chosen place, a random walker needs 345 random steps to arrive at the mosque. The commute time, the expected number of steps a random walker needs to reach the mosque from the church and then to return, equals 405 steps.

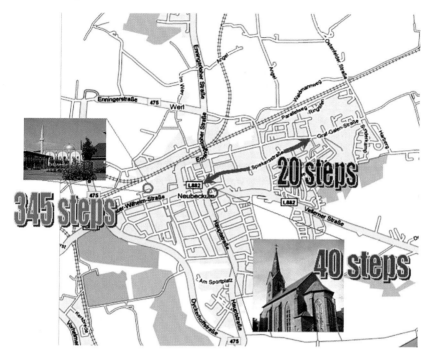

Fig. 2.23 Neubeckum (Westphalia): the church and the mosque in dialog

Spiekersstrasse, the street which is parallel to the railway, now is the best accessible place of motion in Neubeckum playing the role of its structural "center of mass;" it can be achieved from any other location in the city in just 20 random steps. If we estimate relative isolation of other places of motion with respect to Spiekersstrasse by comparing their first-passage times in the logarithmic scale (2.91), then the location of the church is evaluated by $I_{\text{Church}} \approx 3\,\text{dB}$ of isolation, and $I_{\text{Mosque}} \approx 12\,\text{dB}$, for the mosque.

Indeed, isolation was by no means the aim of the Muslim community. The mosque in Neubeckum has been erected on a vacant place, where land is relatively cheap. However, structural isolation under certain conditions would potentially have dramatic social consequences. Efforts to develop systematic dialogue and increased cooperation based on a reinforced culture of consultations are viewed as essential to deliver a sustainable community.

2.7 Summary

We assumed that spatial experience in humans intervening in the city may be organized in the form of a universally acceptable network.

We also assumed that the frequently travelled routes are nothing else but the "projective invariants" of the given layout of streets and squares in the city – the function of its geometrical configuration, which remains invariant whatever origin-destination route is considered.

Based on these two assumptions, we have developed a method that allows us to capture a neighborhood's inaccessibility.

Any finite graph G can be interpreted as a discrete time dynamical system with a finite number of states. The temporal evolution of such a dynamical system is described by a "dynamical law" that maps vertices of the graph into other vertices and can be interpreted as the transition operator of random walks. The level of accessibility of nodes and subgraphs of undirected graphs can be estimated precisely in connection with random walks introduced on them, and We have applied this method to the structural analysis of different cities.

Chapter 3
Exploring Community Structure by Diffusion Processes

The distance travelled is proportional to the time spent travelling, for an ordinary free motion.

Another kind of motion is diffusion describing the spread of any quantity that minimizes its concentration gradient, $\nabla \varrho(\mathbf{r},t)$. It determines the movement of particles from an area where their concentration is high to an area having a lower concentration.

The distance travelled in isotropic space by particles which diffuse as a result of random motion is proportional to the square root of the time spent travelling. The general equation, describing diffusion in $\mathbb{R}^n$ and not referring to any microscopic model, is

$$\frac{\partial \varrho(\mathbf{r},t)}{\partial t} = D\Delta \varrho(\mathbf{r},t), \tag{3.1}$$

where D is the diffusion constant, Δ is the Laplace operator and $\varrho(\mathbf{r},t)$ is the concentration of diffusing material.

While in complex environments, diffusion properties might not be fully characterized by a single diffusion constant being determined by the structure and shapes of the available spaces of motion. Thus, diffusion is described by the Laplace operators satisfying the mean value property,

$$\sum_{i\in V} L_{ij} = 0, \quad \forall j \in V, \tag{3.2}$$

and measured by means of the concentration gradient.

In the this chapter, we use the diffusion process for the purpose of community detection in the spatial graphs of urban environments.

3.1 Laplace Operators and Their Spectra

A passive transport process depends on the permeability of edges and possible delays at nodes. Traffic conditions can be described by an undirected weighted graph, in which a symmetric nonnegative affinity matrix is

$$\mathbf{w} : V \times V \to [0,\infty), \quad w_{ij} = w_{ji},$$

Ph. Blanchard, D. Volchenkov, *Mathematical Analysis of Urban Spatial Networks,*
Understanding Complex Systems, DOI 10.1007/978-3-540-87829-2_3,
© Springer-Verlag Berlin Heidelberg 2009

for each edge $i \sim j$, but $w_{ij} = 0$ for $i \nsim j$. The weight expresses the strength of the flow along the connection.

3.1.1 Random Walks and Diffusions on Weighted Graphs

In models describing lazy random walks, the diagonal elements of the affinity matrix, $0 < w_{ii} \leq 1$, are called the laziness parameters. A lazy random walker located at node $i \in V$ takes edges going out of $i \in V$ with probability

$$\Pr(v_{t+1} = j | v_t = i, \, i \sim j) = \frac{w_{ij}}{\sum_{i \sim j} w_{ij}} > 0, \tag{3.3}$$

but remains in place with probability $1 - w_{ii}$.

In weighted graphs, the degree of a vertex is defined as the sum of weights of all incident edges,

$$k_i \equiv \deg(i) = \sum_{j \sim i} w_{ij}, \tag{3.4}$$

and can be viewed as describing the total traffic through the node.

If we define two diagonal matrices,

$$\mathbf{D} = \mathrm{diag}\left(\{\deg(i)\}_{i \in V}\right), \quad \mathbf{B} = \mathrm{diag}\left(\{w_{ii}\}_{i \in V}\right),$$

then the stochastic operator describing the one-step transitions of a Markov chain on the undirected weighted graph $G(V, E)$ is

$$\mathbf{T}^{(w)} = (\mathbf{1} - \mathbf{B}) + \mathbf{B} \mathbf{D}^{-1} \mathbf{w}. \tag{3.5}$$

The relevant measure for such lazy random walks is

$$\mu^{(w)} = \sum_{j \in V} \frac{\deg(j)}{w_{jj}} \delta(j). \tag{3.6}$$

In particular, if all weights are equal, then the matrix (3.5) turns into the transition matrix of usual random walks (2.27).

The normalized Laplace operator $\mathbf{L}^{(w)}$ associated with the transition matrix (3.5),

$$\mathbf{L}^{(w)} = \mathbf{B}^{-1/2} \mathbf{D}^{-1/2} (\mathbf{1} - \mathbf{T}^{(w)}) \mathbf{D}^{1/2} \mathbf{B}^{1/2}, \tag{3.7}$$

is self-adjoint with respect to the measure (3.6). It follows from (3.7) that if λ is an eigenvalue of $\mathbf{T}^{(w)}$, then $1 - \lambda$ is an eigenvalue of $\mathbf{L}^{(w)}$.

An important special case of (3.7) is the normalized Laplace operator introduced by Chung (1997),

$$\tilde{L}_{ij} = \delta_{ij} - \frac{A_{ij}}{\sqrt{k_i k_j}}. \tag{3.8}$$

Being self-adjoint with respect to the measure (2.23), it describes the diffusion process on undirected graphs. The matrix (3.8) is a symmetric, positive semidefinite matrix associated with the transition matrix of ordinary random walks (2.27).

The symmetric Dirichlet form,

$$\mathscr{D}_w(f) = \left\langle f, L^{(w)} f \right\rangle = f^\star L^{(w)} f = \sum_{i \sim j} w_{ij} \left(f(i) - f(j) \right)^2, \tag{3.9}$$

is defined for all functions $f \in \ell^2 \left(\mu^{(w)} \right)$. Then the Rayleigh-Ritz quotient associated with the Markov process reversible with respect to the measure (3.6) is given by

$$R_\mu(f) = \frac{\mathscr{D}_w(f)}{\|f\|^2}, \text{ for } f \neq 0. \tag{3.10}$$

If f is an eigenvector of $\mathbf{L}^{(w)}$ belonging to the eigenvalue λ, then $R_\mu(f) = \lambda$.

The Dirichlet form induces a seminorm $\|f\|_{L^{(w)}} = \left\langle f, L^{(w)} f \right\rangle$ on $\ell^2(\mu^{(w)})$ and quantifies how much the function $f \in \ell^2(\mu^{(w)})$ varies locally. It is well-known (see Chung 1997) that the minimization of the symmetric Dirichlet form (3.9) with the additional constraint of $\|f\|_{L^{(w)}} = 1$ is equivalent to finding the eigenvectors of the matrix $\mathbf{D}^{-1/2} \mathbf{w} \mathbf{D}^{1/2}$ where $\mathbf{w}$ is the symmetric affinity matrix.

The Rayleigh-Ritz quotient (3.10) is the basic ingredient for the standard eigenvalue algorithms (see Trefethen et al. 1997), in which one first minimizes the Rayleigh-Ritz quotient over the whole vector space. It gives the lowest eigenvalue, which is always $\lambda_1 = 0$, for the Laplace operator due to the mean value property (3.2). Next, one restricts attention to the orthogonal complement of the major eigenvector found in the first step and minimizes over this subspace. That produces the next lowest eigenvalue and corresponding eigenvector. One can continue this process recursively. At each step, one minimizes the Rayleigh-Ritz quotient over the subspace orthogonal to all the vectors found in the preceding steps to find another eigenvalue and its corresponding eigenvector.

In particular, the eigenvalues of the normalized Laplace operator (3.8) defined on a connected graph satisfy

$$0 = \lambda_1 \leq \ldots \leq \lambda_N \leq 2, \tag{3.11}$$

(see Chung (1997)).

3.1.2 Diffusion Equation and its Solution

The equation governing the diffusion process defined on undirected graphs is a special case of a general linear dynamical system describing the expectation number $\mathbf{n}(t) \in V \times \mathbb{Z}_+$ of random walkers

$$\dot{\mathbf{n}} = \mathbf{Ln}, \tag{3.12}$$

in which the Laplace operator has a discrete spectrum. The equation is supplied with an initial condition $\mathbf{n}_0$.

Being defined on undirected graphs, the operator $\mathbf{L}$ is normal, $\mathbf{LL}^T = \mathbf{L}^T\mathbf{L}$ and, consequently, has a complete set of orthogonal eigenvectors. If the graph does not change with time, the dynamical system (3.12) is autonomous and the solution is

$$\mathbf{n}(t) = \exp(\mathbf{L}t)\,\mathbf{n}_0, \tag{3.13}$$

if all eigenvalues are simple. While in general it is solved by the function $\mathbf{n}^t = \mathbf{Q}\exp(-t\mathbf{J})(\mathbf{Q}^{-1}\mathbf{n}_0)$, in which $\mathbf{J}$ is a block diagonal matrix (the Jordan canonical form of $\mathbf{L}$), and $\mathbf{Q}$ is the transformation matrix corresponding to the Jordan form. Given the eigenvalue λ_α with the multiplicity $m_\alpha > 1$, the relevant contribution to $\mathbf{n}^t$ is given by

$$\exp(-t\lambda_\alpha)\sum_{k=0}^{m_\alpha}\left(\sum_{l=0}^{m_\alpha-1}c_{l+1}\frac{t^l}{l!}\right)\mathbf{u}_k \tag{3.14}$$

where $\mathbf{u}_k$ is the k-th vector of the m_α-dimensional subspace belonging to the degenerate eigenvalue λ_α and c_l is the l-th component of the transformed vector of initial conditions, $\mathbf{c} = \mathbf{Q}^{-1}\mathbf{n}_0$. It follows from (3.14) that in the presence of degenerate modes with $m_\alpha \gg 1$ the relaxation process can be significantly delayed.

3.1.3 Spectra of Special Graphs and Cities

In Table 3.1, we have presented the spectra of the normalized Laplace operators defined on special graphs along with their multiplicities: the complete graph $\mathbb{K}_N$ on N vertices; the cycle $\mathbb{C}_N$ on N vertices; the path $\mathbb{P}_N$ on N vertices (a linear chain); the d-dimensional hypercube $\mathbb{H}_d$ on $N = 2^d$ nodes; the star graph $\mathbb{S}_N$ having one pivotal node (a hub) and $N - 1$ twin nodes connected to it. The details of calculations can be found in Chung (1997).

The spectra shown in Table 3.1 bring out the symmetries of the special graphs. For instance, the spectrum of the normalized Laplace operator (3.8) defined on a cycle $\mathbb{C}_N$, $\lambda_s = 1 - \cos(\pi s/N)$ is symmetric with respect to the transformation of parameter, $s \to N - s$: $\lambda_s = \lambda_{N-s}$, $s = 0,\ldots N - 1$ and, therefore, each eigenvalue in the spectrum is effectively doubled reflecting the symmetry of cycles with respect to the cyclic permutation of its vertices (deVerdiere 1998). Permutations of nodes in complete graphs and of the twin nodes in star graphs do not affect the diffusion process, so that the eigenvalues of the Laplace operator defined on these graphs exhibit the highest multiplicity. The spectrum of d-dimensional hypercubes unveils the d-dimensional reflection symmetry.

It is not a surprise that the spectra of the normalized Laplace operator (3.8) defined on the spatial graphs of compact urban patterns–the spectra of cities-are totally different from those presented in Table 3.1.

Table 3.1 The spectra of the normalized Laplace operators defined on the special graphs

Graph	Spectrum	Multiplicity
$\mathbb{K}_N$	$0,$ $N/(N-1)$	1 $N-1$
$\mathbb{C}_N$	$1 - \cos(2\pi s/N),$ $s = 0,\ldots,N-1$	$1(2)$
$\mathbb{P}_N$	$1 - \cos(\pi s/(N-1)),$ $s = 0,\ldots,N-1$	1
$\mathbb{H}_N$	$2s/d,$ $s = 0,\ldots d.$	$\binom{d}{s}$
$\mathbb{S}_N$	0 1 2	1 $N-2$ 1

If we take many, many random numbers from an interval of all real numbers symmetric with respect to a unit and calculate the sample mean in each case, then the distribution of these sample means will be approximately normal in shape and centered at 1 provided the sample size was large. The probability density function of a normal distribution forms a symmetrical bell-shaped curve highest at the mean value indicating that in a random selection of numbers around the mean (1) have there is a higher probability of being selected than those far away from the mean. Maximizing information entropy among all distributions with known mean and variance, the normal distribution arises in many areas of statistics.

It is interesting to compare the empirical distributions of eigenvalues over the interval $[0, 2]$ in the city spectra with the normal distribution centered at 1. In Fig. 3.1, we have shown a probability-probability plot of the normal distribution (on the horizontal axis) against the empirical distribution of eigenvalues in the city spectra (the normal plot). A random sample of the normal distribution, having size equal to the number of eigenvalues in the spectrum has been be generated, sorted ascendingly, and plotted against the response of the empirical distribution of city eigenvalues.

The spectra of all organic cities – the downtown of Bielefeld and the Rothenburg o.d.T. (see Fig. 3.1) are the examples – are akin to the Gaussian curve centered at 1.

Figure 3.2 shows the normal plot analogous to that given in Fig. 3.1, but for the empirical distribution of eigenvalues in the spectra of the city canal networks in Venice and Amsterdam. The spectra of canals maintained in the compact urban patterns of Venice and Amsterdam also look amazingly alike and are obviously tied to the normal distribution, although these canals had been founded in the dissimilar geographical regions and for different purposes. While the Venetian canals mostly serve the function of transportation routes between the distinct districts of the gradually growing naval capital of the Mediterranean region, the concentric web of Amsterdam gratchen had been built in order to defend the city.

In order to compare these two spectra in detail, we have implemented the method of kernel density estimations (or the window method Parzen 1962). Kernel density estimation works by considering the location of each data point and replacing that data point with a kernel function which has an area of one (see Wasserman 2005).

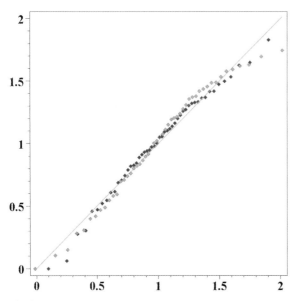

Fig. 3.1 The probability-probability plot of the normal distribution (on the horizontal axis) against the empirical distribution of eigenvalues in the city spectra of German medieval cities, Bielefeld and Rothenburg o.d.T. The diagonal line $y = x$ is set for a reference

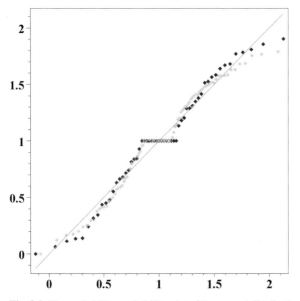

Fig. 3.2 The probability-probability plot of the normal distribution (on the horizontal axis) against the empirical distribution of eigenvalues in the spectra of the city canal networks in Venice and Amsterdam. The diagonal line $y = x$ is set for a reference

The kernels are then individually summed up over all data points and normalized, so that the estimate is a probability density function.

In Fig. 3.3, we have compared the spectral density distributions for the city canal networks in Venice and Amsterdam. The graph shown in Fig. 3.3 represents the result of an attempt to perform kernel density estimation on the sets of eigenvalues in order to develop an approximation to the probability density function that these spectra could have been drawn from. The smooth Gaussian kernel has been used in developing the estimate displayed in Fig. 3.3. While performing the discrete case of kernel density estimation, we have employed the discrete Fourier transform in order to facilitate calculations of the kernel sums. This results in an estimating function which is periodic over the given range. In principle, the application of the discrete Fourier transform indicate that estimates near the lower and upper boundaries of the range will often be more imprecise than points within the range. However, for the analysis of city spectra this effect is of no importance since the distributions of eigenvalues observed for the city spatial graphs are usually centered in the middle of the range.

Probably, the most interesting observation related to city spectra is the accumulation of eigenvalues at the center of range. Moreover, the spectral density distributions presented in Fig. 3.3 indicate that both spectra of city canal networks exhibit a "moderate" degeneracy of the eigenvalue $\lambda = 1$ of the normalized Laplace operator (3.8) that is equivalent to the zero-eigenmode $\mu = 0$ of the transition operator of random walks which we have discussed in the previous chapter. The high multiplicity of eigenvalues, in particular of the eigenvalue $\lambda = 1$ ($\mu = 0$, for random walks), is a fascinating feature of graph spectra related to transport networks (see Fig. 3.4).

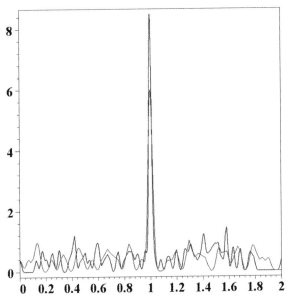

Fig. 3.3 The spectral density distributions for the city canal networks in Venice and Amsterdam estimated by their kernel densities

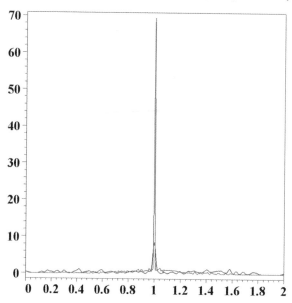

Fig. 3.4 The spectral density distribution for the city canal network in Venice is compared with that of the street grid array in Manhattan. Both spectral density distributions have been estimated by their kernel densities

It is remarkable that the spectral density distributions shown in Figs. 3.3 and 3.4 are dramatically dissimilar to those reported for the random graphs of Erdös and Rényi studied by Farkas et al. (2001, 2002).

The classical Wigner semicircle distribution (see Abramowitz et al. 1972) arises as the limiting distribution of eigenvalues of many random symmetric matrices as the size of the matrix approaches infinity (Sinai et al. 1998). In accordance with it, in random graphs, the nontrivial eigenvalues of their adjacency matrices cluster close to 1 (Arnold 1971, Alon et al. 2002). This fact remains true even for the spectra of scale-free random graphs with a power-law degree distribution that has been observed for the scale-free random tree-like graphs by Dorogovtsev et al. (2003). In Eriksen et al. (2003), the density distributions of eigenvalues for the Internet graph on the Autonomous Systems level had been presented. These distributions appear to be broad and have two symmetric maxima being similar to the spectral density distribution reported for random scale-free networks. The eigenvalues of the normalized Laplace operator in a random power-law graph also follow the semicircle law (Chung et al. 2003), whereas the spectrum of the adjacency matrix of a power-law graph obeys the power law (Bollobás et al. 2002).

In contrast, the spectral density distributions for compact urban patterns are either bell-shaped, or have a sharp peak at $\lambda = 1$. City spectra reveal the profound structural dissimilarity between urban networks and networks of other types studied before. This multiple eigenvalue appears due to twin nodes in the spatial graphs. In the previous chapter we discussed that these twin nodes always come out when some geometrical motif is repeatedly presented in the urban environment.

Twins would arrive as the cliental nodes of star graphs being connected to one and the same hub – a star graph could represent the urban sprawl developments. They also can be found in the complete bipartite subgraphs that encode in spatial graphs of the ideally regular street grids. These structures appear to be overrepresented in some compact urban patterns.

In Sect. 3.1.5, we shall see that the accumulation of eigenvalues close to 1 indicates that the graph has good expander property, i.e., it constitutes the economical robust network in which the number of edges growing approximately linearly with size, for all subsets.

The multiple eigenvalue $\lambda = 1$ would score a valuable fraction of all eigenvalues for the spatial graphs of modern cites – up to 48 percent of all eigenvalues in the spectrum of Manhattan is $\lambda = 1$ (see Fig. 3.4). In this case, the resulting distribution of eigenvalues can essentially depart from any standard Gaussian curve of normal distributions (see the normal plot presented in Fig. 3.5).

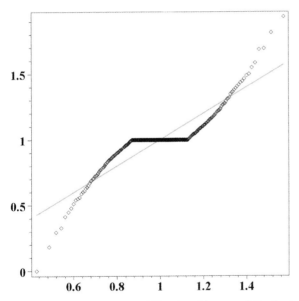

Fig. 3.5 The probability-probability plot of the normal distribution (on the horizontal axis) against the empirical distribution of eigenvalues of the spatial graph of Manhattan. The diagonal line $y = x$ is set for a reference

3.1.4 Cheeger's Inequalities and Spectral Gaps

Given a stationary random walk of the nearest neighbor type defined on an undirected weighted graph G specified by some measure m, for any subset $\Gamma \subset G$ we define the probability

$$\eta(\Gamma) = \Pr\left[v_{t+1} \in G \setminus \Gamma \mid v_t \in \Gamma \subset G\right] \tag{3.15}$$

that a random walker located somewhere in the subgraph $\Gamma \subset G$ moves out to its complement $G \setminus \Gamma$ in the next step. It seems natural to use this probability for estimating transport property of the entire network.

If we denote the edge boundary of the subgraph $\Gamma \subset G$ by $\partial \Gamma = \{i \sim j, i \in \Gamma, j \in G \setminus \Gamma\}$, then it is easy to see that

$$\eta(\Gamma) = \Pr[v_{t+1} \in G \setminus \Gamma | v_t \in \Gamma] = \frac{|\partial \Gamma|}{m(\Gamma)} \tag{3.16}$$

where $m(\Gamma) = \sum_{i \in \Gamma} m(i)$ is the total volume of Γ with respect to m. For instance, if we take the counting measure m_0 defined by (2.17), then $m_0(G) = N$ and $m_0(\Gamma) = |V_\Gamma|$ are the number of nodes in G and in its subgraph $\Gamma \subset G$. Otherwise, if we use the measure (2.23) associated with random walks, then $m(G) = 2M$ is the total number of edges in the graph, and $m(\Gamma) = \sum_{i \in \Gamma} \deg(i)$.

For subgraphs Γ such that $m(\Gamma) \leq m(G \setminus \Gamma)$, the probability $\eta(\Gamma)$ defined by (3.16) is called the Cheeger ratio for the subgraph Γ. It has been introduced by Cheeger (1969) and is used with optimal bisections of graphs by spectral methods. In particular, the minimal value of (3.16) among all Cheeger ratios calculated for all possible subgraphs Γ such that $m(\Gamma) \leq m(G \setminus \Gamma)$ is called the Cheeger constant (or Cheeger's number) of the graph G,

$$h_G = \min_{\{\Gamma : m(\Gamma) \leq m(G \setminus \Gamma)\}} \eta(\Gamma). \tag{3.17}$$

If we want to cut a small subset Γ from the rest of G, we have to sever at least $h_G m(\Gamma)$ edges.

The Cheeger constant (3.17) is a numerical measure of whether or not a graph has a "bottleneck" (see Donetti et al. 2006, Lackenby 2006). Its value is strictly positive if and only if G is a connected graph. It seems intuitive that if the Cheeger constant is small but positive, then a "bottleneck" exists in the sense that there are two "large" sets of vertices with "few" links (edges) between them. Finally, the Cheeger constant is large if any possible division of the vertex set into two subsets has "many" links between those two subsets.

However, in practical applications the Cheeger constant is very difficult to compute directly following (3.17), especially if the graph G is large since it is required to analyze approximately 2^{N-1} subsets of graph vertices. The spectral methods allow us to estimate the Cheeger constant by the second largest eigenvalue of the transition matrix of random walks (as it had been done by Cheeger (1969) himself) or, alternatively, by the second smallest eigenvalue of the normalized Laplace operator, as Chung (1997) did.

Given a connected, undirected graph G, let $0 = \lambda_1 < \lambda_2 \leq \ldots \leq \lambda_N$ be the ordered eigenvalues of the normalized Laplace operator (3.8) and h_G be the Cheeger constant of G. Then, accordingly to both Cheeger (1969) and Chung (1997), the following inequalities hold:

$$2h_G \geq \lambda_2 \geq \frac{h_G^2}{2}. \tag{3.18}$$

It follows from (3.18) that the second smallest eigenvalue of the normalized Laplace operator (which can be called the spectral gap since always $\lambda_1 = 0$) contains a lot of information about the graph structure.

In order to give an idea about the typical values of spectral gaps for the city spatial graphs, we have ordered them in Fig. 3.6 with respect to their sizes. The values of spectral gaps in the spatial graphs of compact urban patterns are indicated by the radiuses of the blobs. The vertical position of each blob is determined by the size of the relevant spatial graph.

The inequalities (3.18) give the minimal and maximal bounds for the value of Cheeger's constant, but do not determine it exactly. In order to show how typical the values of these bounds are, in Fig. 3.7 we have compared them with the averaged maximal bounds derived for the connected random graphs $G(N, M)$ which have the same numbers of nodes and edges as the original city spatial graphs.

Random graphs are widely used in the probabilistic method, where one tries to prove the existence of graphs with certain properties (Bollobás et al. (2002)). We have constructed the ensembles of the connected realization of random graphs $G(N, M)$. Despite the fact that random graphs had about the same number of edges and nodes as the actual city spatial graphs have, they missed the intrinsic topological structure of urban patterns. While calculating the maximal bounds, we have averaged the spectral gaps over ensembles of different connected realizations of $G(N, M)$.

The relatively small values of bounds for the Cheeger constants observed for the actual city spatial graphs indicate a possible presence of bottlenecks in city fabric. The maximal bounds for Cheeger's constant obtained for the random graphs are higher, since their edges are distributed identically random between nodes and, therefore, the appearance of bottlenecks in random graphs is less probable.

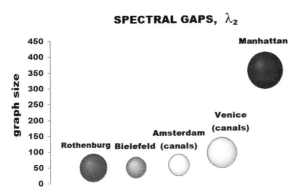

Fig. 3.6 The values of spectral gaps in the spatial graphs for the compact urban patterns are shown by the radiuses of bulbs: the downtown of Bielefeld ($\lambda_2 = 0.06058$), Rothenburg ob der Tauber ($\lambda_2 = 0.10568$), the network of canals in Venice($\lambda_2 = 0.12542$) and in Amsterdam($\lambda_2 = 0.06427$), in Manhattan ($\lambda_2 = 0.18352$). On the left side, we have presented the scale indicating the sizes of the correspondent spatial graphs

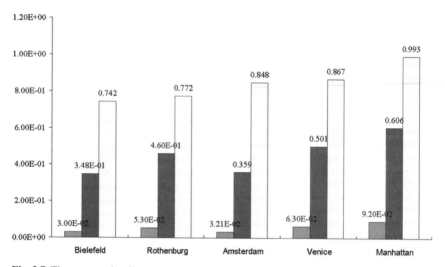

Fig. 3.7 The comparative diagram represents the minimal and maximal bounds for Cheeger's constants for the city spatial graphs and their random counterparts. Left bars indicate the minimal bounds for the Cheeger constant for cities, central bars stay for the maximal bounds, and right bars give the averaged maximal bounds for the Cheeger constant estimated for the connected random graphs $G(N, M)$ that have the same numbers of nodes and edges as the original city spatial graphs

3.1.5 Is the City an Expander Graph?

An expander graph is a sparse graph which has high connectivity properties (Chung et al. 1989b), – any two disjoint sets of vertices cannot be disconnected from each other without removing a lot of edges. A random walk defined on an expander graph converges very rapidly to the uniform distribution (Linial et al. 2005). Expander graphs are useful for a broad spectrum of applications in computer science, from the design of good routing networks to de-randomization. The original motivation for expanders was to build economical robust networks: an expander with bounded valence is precisely an asymptotic robust graph with a number of edges that grows linearly with size, uniformly for all its subsets.

A random d-regular graph has good expansion, with high probability. The expander mixing lemma (Linial et al. 2005) states that, for any two subsets $\Gamma_1 \subset G$ and $\Gamma_2 \subset G$ of a regular expander graph G, the number of edges between Γ_1 and Γ_2 is approximately what we expect for a random d-regular graph,

$$e_d(\Gamma_1, \Gamma_2) \simeq d \cdot \frac{\deg(\Gamma_1)\deg(\Gamma_2)}{N},$$

where $\deg(\Gamma_{1,2}) = \sum_{i \in \Gamma_{1,2}} \deg(i)$.

In order to quantify the possible dissimilarity between the city spatial graphs and expanders, we use the standard discrepancy indicator, $\mathrm{disc}(G)$, expressing how the edges are distributed between sets in the network graph G and its random counterpart (see Chung 1997, Butler 2006). From their works, it is known that the expected number of edges that should connect to different subgraphs, Γ_1 and Γ_2, in a random graph model with uneven degree distributions is equal to $\deg(\Gamma_1)\deg(\Gamma_2)/2M$. Then, the discrepancy index $\mathrm{disc}(G)$ can be defined as the minimal α in the following estimation:

$$\left| e(\Gamma_1, \Gamma_2) - \frac{\deg(\Gamma_1)\deg(\Gamma_2)}{2M} \right| \leq \alpha \sqrt{\deg(\Gamma_1)\deg(\Gamma_2)} \qquad (3.19)$$

where $\sqrt{\deg(\Gamma_1)\deg(\Gamma_2)}$ is the error normalizing term. The value of $\mathrm{disc}(G) \to 0$ if the structure of graph G is close to random. If the discrepancy α is small, then the graph has a good expander property.

The discrepancy of the graph is also related to the spectrum of the normalized Laplace operator (Chung 1997). Given $0 = \lambda_1 \leq \lambda_2 \leq \ldots \leq \lambda_N$, the ordered set of eigenvalues of the normalized Laplace operator (3.8), then it has been proven by Butler (2006) that the following inequality holds:

$$\begin{aligned} \mathrm{disc}(G) &\leq \max\{|1 - \lambda_2|, |\lambda_N - 1|\} \\ &\leq 150\,\mathrm{disc}(G)(1 - 8\log(\mathrm{disc}(G))). \end{aligned} \qquad (3.20)$$

With regard to (3.20), the graph discrepancy index $\mathrm{disc}(G)$ bounds the maximal deviations of eigenvalues of the normalized Laplace operator from 1. If the value of $\mathrm{disc}(G)$ is small, the eigenvalues are clustered close to $\lambda = 1$, while they could be almost normally distributed over the interval $[0, 2]$ otherwise. The spectral density distributions for the city spatial graphs which we have discussed above indicate that most of them have very good expander property. Expander graphs are known for solving the problem of efficient network design by maximizing the flow of information through it while minimizing the cost of building the network. Therefore, the spatial graphs encoding the transport system of the city appear to be well established and look optimal.

The inequalities (3.20) derived by Butler (2006) determine only the bounds for discrepancy. The minimal bound for them is zero, but the maximal bound, although very small can vary from city to city (see Fig. 3.8). The small discrepancy values observed for the city spatial graphs indicate that the distributions of edges illuminating the adjacency relations in urban environments are essentially close to those in random regular graphs. However, as we have seen before, in contrast to random graphs, the actual city spatial graphs probably have bottlenecks, or more generally, can consist of few relatively weakly connected components that are not typical for random graphs at all.

We discuss the segmentation of city spatial graphs into components in the next section.

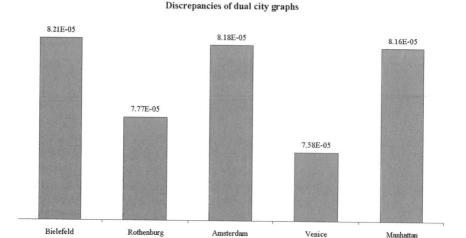

Fig. 3.8 The maximal bounds for the discrepancy values of dual city graphs

3.2 Component Analysis of Transport Networks

Component analysis is an important problem in transport network studies stipulating the creation of service districts based on accessibility, the drive-time analysis, and evaluating the best facility locations. Therefore, the efficient partitioning of the transport network is one of the most effective logistics optimization tools and the biggest opportunity for most companies to significantly reduce costs.

The component analysis is also of importance when dealing with extremely large graphs, when we need to cluster the vertices into logical components for storage (to improve virtual memory performance) or for drawing purposes (to collapse dense subgraphs into single nodes in order to reduce cluttering). Various graph partitioning algorithms have arisen in computer science as a preprocessing step to divide-and-conquer algorithms, which work by recursively breaking down a problem into two or more subproblems of the same (or related) type (Frigo et al. 1999).

3.2.1 Graph Cut Problems

In many transport networks, their individual components are sparsely connected by only a few "bridges" between them.

There is extensive literature on clustering and partitioning of graphs in two or more almost disjoint parts. Several different flavors of graph partitioning can be implemented depending on the desired objective function. The smallest set of edges to cut (the minimum cut set) that will disconnect a graph can be efficiently found using network flow methods described by Ahuja et al. (1993).

A better partition criterion seeks a small cut that partitions the vertices into roughly equalsized pieces. The basic approach to dealing with graph partitioning

or max-cut problems is to construct an initial partition of the vertices (either randomly or according to some problem-specific strategy) and then sweep through the
vertices, deciding whether the size of the cut would increase or decrease if we moved
this vertex over to the other side.

Graph cut problems are often NP-hard (Nondeterministic Polynomial time hard).
Let us assume that $G(V,E)$ is a connected undirected weighted graph with a symmetric affinity matrix $w_{ij} = w_{ji}$, if $i \sim j$, and $w_{ij} = 0$ otherwise. The degree $\deg(i)$
of a node is defined as the sum of all weights of incident edges (3.4). In order
to achieve a balanced partition of G in two components, we can use the Cheeger
constant (3.17) of the graph G introduced in the Sect. 3.1.4. For the optimal partition $G = \Gamma \cup \{G \setminus \Gamma\}$ (at which the minimum conductance between two partitions
is achieved) the value of the relevant Cheeger ratio (3.16) coincides with Cheeger's
constant,

$$h_G = \min_{\Gamma \subset G} \frac{|\partial \Gamma|}{\min\left(m(\Gamma), m(G \setminus \Gamma)\right)} \tag{3.21}$$

where the volume of the subgraph Γ is denoted by $m(\Gamma)$ and the volume of the edge
boundary $\partial \Gamma$ connecting the subgraph Γ to its complement $G \setminus \Gamma$ is defined by

$$|\partial \Gamma| = \sum_{i \in \Gamma, \, j \in G \setminus \Gamma} w_{ij}. \tag{3.22}$$

Another objective function known as the normalized cut is defined by

$$n(G) = \min_{\emptyset \subset \Gamma \subset G} \left(\frac{|\partial \Gamma|}{m(\Gamma)} + \frac{|\partial \Gamma|}{m(G \setminus \Gamma)} \right) \tag{3.23}$$

has been proposed by Shi et al. (2000).

It is obvious that both objective functions, (3.22) and (3.23), are closely related,
since for any pair of positive numbers $a, b > 0$ it is always true that

$$\min(a,b) \leq \left(\frac{1}{a} + \frac{1}{b} \right)^{-1} \leq 2\min(a,b).$$

In particular, both cut problems are NP-hard, because of a direct checking of all
2^{N-1} possible subsets of G is practically impossible especially if the graph is large,
so that the graph segmentation problem needs to be approximated by computationally feasible methods.

A spectral heuristic can be implemented in order to detect bottlenecks and weakly
connected subgraphs.

3.2.2 Weakly Connected Graph Components

A good heuristic can be obtained by writing the objective function as a quadratic
form (by the Rayleigh quotient) and relaxing the discrete optimization problem to a
continuous one which then can be solved using the standard methods.

Let us suppose that the connected undirected weighted graph G contains two components, G_1 and G_2, with just a few edges

$$e(G_1, G_2) = \{i \sim j,\ i \in G_1,\ j \in G_2\}$$

linking them. Transport property of a network spanned by the graph G can be characterized by the following reciprocal probabilities,

$$
\begin{aligned}
P_{G_1 G_2} &= \Pr[G_1 \to G_2 | G_1] = m^{-1}(G_1) \Sigma_{i \in G_1,\ j \in G_2} w_{ij}, \\
P_{G_2 G_1} &= \Pr[G_2 \to G_1 | G_2] = m^{-1}(G_2) \Sigma_{i \in G_2,\ j \in G_1} w_{ij},
\end{aligned}
\tag{3.24}
$$

that a random walker located in one of the components alternates it in the next step.

In a probabilistic setting, we say that two components, G_1 and G_2, of the graph $G = G_1 \cup G_2$ are the weakly connected components if the probability

$$\Pr[G_1 \leftrightarrow G_2] = \left(m^{-1}(G_1) + m^{-1}(G_2)\right) \sum_{i \in G_1,\ j \in G_2} w_{ij}, \tag{3.25}$$

of random traffic between them in one-step is ever minimal among all possible bisections of the graph G.

The minimization of the inter-subgraph random traffic probability can be promptly reformulated as a discrete optimization problem (von Luxburg et al. 2004). Let us define the indicator vector χ_i for the subgraphs G_1 and G_2 as

$$\chi_i = \begin{cases} 1,\ i \in G_1, \\ -1,\ i \in G_2 \end{cases} \tag{3.26}$$

and note that $\chi^\top \chi = |V_G|$. Then, it can be readily obtained that

$$
\begin{aligned}
\chi^\top \mathbf{L} \chi &= \Sigma_{[i \sim j]} w_{ij} (\chi_i - \chi_j)^2 = 4|e(G_1, G_2)|, \\
\chi^\top \mathbf{D} \chi &= \Sigma_{i \in G} w_{ij} = \mu(G_1) + \mu(G_2) = \mu(G),
\end{aligned}
\tag{3.27}
$$

where $\mathbf{D} = \operatorname{diag}\{\deg(i), i \in V\}$ is the diagonal matrix of vertex degrees and $\mathbf{L} = \mathbf{D} - w$ is the canonical Laplace operator defined on the weighted graph G. If we then denote by $\mathbf{e}$ the column vector of all ones, we can write

$$
\begin{aligned}
\chi^\top \mathbf{D} \mathbf{e} &= \Sigma_{i \in G_1} \deg(i) - \Sigma_{i \in G_2} \deg(i) \\
&= \mu(G_1) - \mu(G_2),
\end{aligned}
\tag{3.28}
$$

so that $\chi^\top \mathbf{D} \mathbf{e} = 0$ if and only if the components G_1 and G_2 are the volume-balanced components. It is then obvious that

$$\min_{G_1 \cup G_2 = G} \Pr[G_1 \leftrightarrow G_2] = \frac{\mu(G)}{4} \cdot \min_{f \mathbf{D} \mathbf{e} = 0} \frac{f^\top \mathbf{L} f}{f^\top \mathbf{D} f} \tag{3.29}$$

where the r.h.s. Rayleigh quotient is minimized over all vectors $f \in \{-1, 1\}^N$. If instead, we suppose that f can take real values, $f \in \mathbb{R}^N$ (von Luxburg et al. 2004), then standard linear algebra arguments show that the minimal value of (3.29),

$$\lambda_2 = \min_{f\mathbf{De}=\mathbf{0}} \frac{f^{\top}\mathbf{L}f}{f^{\top}\mathbf{D}f}, \tag{3.30}$$

is achieved for the eigenvector f_2 (the Fiedler eigenvector, studied by Fiedler (1975)) which belongs to the second smallest eigenvalue λ_2 of the generalized eigenvector problem,

$$\mathbf{L}f_2 = \lambda_2\mathbf{D}f_2. \tag{3.31}$$

While the smallest eigenvalue of the Laplace operator is always $\lambda_1 = 0$, the second smallest eigenvalue $\lambda_2 > 0$ if the graph G is connected. The normalized spectral clustering (von Luxburg et al. 2004) is given by the Fiedler eigenvector,

$$G_1 = \{i : f_2(i) > 0\}, \quad G_2 = \{i : f_2(i) \leq 0\}. \tag{3.32}$$

In general, each nodal domain on which the components of the αth smallest eigenvector f_α do not change refers to a coherent flow of random walkers (characterized by its decay time $\tau_\alpha \simeq -1/\lambda_\alpha$) toward the domain of the alternative sign. Nodal domains participating in the different diffusion eigenmodes as one and the same degrees of freedom can be considered dynamically independent modules of a transport network (Volchenkov et al. 2007a). It is known from Davis et al. (2001) that the eigenvector f_α can have, at most, $\alpha + m_\alpha - 1$ strong nodal domains (the maximal connected induced subgraphs on which the components of eigenvectors have a definite sign) where m_α is the multiplicity of the eigenvalue λ_α, but not less than 2 strong nodal domains (for $\alpha > 1$) (Biyikoğlu et al. 2004). However, the actual number of nodal domains can be much smaller than the bound obtained in Biyikoğlu et al. (2004). In the case of degenerate eigenvalues of the Laplace operator, the situation becomes even more difficult because this number may vary considerably depending upon which vector from the m_α-dimensional subspace belonging to the degenerate eigenvalue λ_α is chosen.

3.2.3 Graph Partitioning Objectives as Trace Optimization Problems

The generalized graph partitioning problem seeks to partition a weighted undirected graph G into n almost disjoint clusters $\Gamma_1, \ldots \Gamma_n$ such that $\bigcup_{i=1}^{n} \Gamma_n \subset G$ and either their properties share some common trait, or the graph nodes belonging to them are close to each other according to some distance measure defined for the nodes of the graph.

A number of different graph partitioning strategies for undirected weighted graphs have been studied in connection with Object Recognition and Learning in Computer Vision (see Morris (2004)).

Below, we generalize them into four basic cases:

- The Ratio Cut objective seeks to minimize the cut between clusters and the remaining vertices (Chan et al. 1994),

$$\text{Rcut}(G) = \min_{\{\Gamma_v\}} \sum_{v=1}^{n} \frac{|\partial \Gamma_v|}{|V_{\Gamma_v}|}, \tag{3.33}$$

where

$$|\partial \Gamma_v| = \sum_{i \in \Gamma_v, \, j \in G \backslash \Gamma_v} w_{ij}$$

is the size of boundary $\partial \Gamma_v$, $w_{ij} > 0$ is a symmetric matrix of edge weights, $|V_{\Gamma_v}|$ is the number of nodes in the subgraph Γ_v.

- The Normalized Cut objective (3.23) mentioned at the beginning of this section is one of the most popular graph partitioning objectives (Shi et al. 2000, Yu et al. 2003) that seeks to minimize the cut relative to the size $\mu(\Gamma_v)$ of a cluster Γ_v with respect to some counting measure μ instead of the number of its nodes $|V_{\Gamma_v}|$ used in (3.33).
  In particular, given the measure (2.23) associated to random walks then $\mu(\Gamma_v) = \sum_{i \in \Gamma_v} \deg(i)$,

$$\text{Ncut}(G) = \min_{\{\Gamma_v\}} \sum_{v=1}^{n} \frac{|\partial \Gamma_v|}{\mu(\Gamma_v)}. \tag{3.34}$$

- The Ratio Association objective (also called average association) (Shi et al. 2000) aims to maximize the size of a cluster,

$$|\Gamma_v| = \sum_{i,j \in \Gamma_v} w_{ij},$$

relative to the number of its nodes $|V_{\Gamma_v}|$,

$$\text{RAssoc}(G) = \max_{\{\Gamma_v\}} \sum_{v=1}^{n} \frac{|\Gamma_v|}{|V_{\Gamma_v}|}. \tag{3.35}$$

- The Weighted Ratio Association objective (Dhillon et al. 2004a) generalizes (3.35) for the size $\mu(\Gamma_v)$ of the cluster with respect to the measure μ:

$$\text{WRAssoc}(G) = \max_{\{\Gamma_v\}} \sum_{v=1}^{n} \frac{|\Gamma_v|}{\mu(\Gamma_v)}. \tag{3.36}$$

All graph partition objectives (3.34) (3.35) and (3.36) can be formulated as trace maximization problems (Dhillon et al. 2004a). We introduce partition indicator vectors $\{\chi_v\}_{v=1}^{n} \in \{0,1\}^N$ by

$$\chi_v(i) = \begin{cases} 1, & i \in \Gamma_v, \\ 0, & i \notin \Gamma_v \end{cases} \tag{3.37}$$

and note that $\chi_v^\top \chi_v = |\Gamma_v|$.

The Ratio Association objective (3.35) can be re-formulated as

$$\text{RAssoc}(G) = \max\left\{\sum_{v=1}^{n} \frac{\chi_v^{\top} \mathbf{w}\chi_v}{\chi_v^{\top}\chi_v}\right\}$$

$$= \max_{\mathbf{X}^{\top}\mathbf{X}=\mathbf{1}} \text{tr}(\mathbf{X}^{\top}\mathbf{w}\mathbf{X}), \tag{3.38}$$

where $\mathbf{w}$ is the graph affinity matrix, and

$$\mathbf{X} = \left[\frac{\chi_1}{\sqrt{|\Gamma_1|}}, \dots, \frac{\chi_n}{\sqrt{|\Gamma_n|}}\right] \tag{3.39}$$

is the $n \times N$ rectangular orthogonal matrix ($\mathbf{X}^{\top}\mathbf{X} = \mathbf{1}_n$) of normalized indicator vectors.

The Ratio Cut objective (3.33) can be rewritten in terms of the canonical Laplace operator defined on the weighted graph G,

$$\text{Rcut}(G) = \max_{\mathbf{X}^{\top}\mathbf{X}=\mathbf{1}_n} \text{tr}(\mathbf{X}^{\top}\mathbf{L}\mathbf{X}), \tag{3.40}$$

in which $\mathbf{L} = \mathbf{D} - \mathbf{w}$ and $\mathbf{D} = \text{diag}\{\deg(i), i \in V\}$ is the diagonal matrix of vertex degrees, $\deg(i) = \sum_{i \sim j} w_{ij}$.

The Normalized Cut objective can be formulated as

$$\text{Ncut}(G) = \max\left\{\sum_{v=1}^{n} \frac{\chi_v^{\top}\mathbf{w}\chi_v}{\chi_v^{\top}\mathbf{D}\chi_v}\right\}$$

$$= \max_{\tilde{\mathbf{X}}^{\top}\tilde{\mathbf{X}}=\mathbf{1}_n} \text{tr}(\tilde{\mathbf{X}}^{\top}\mathbf{D}^{-1/2}\mathbf{w}\mathbf{D}^{-1/2}\tilde{\mathbf{X}}), \tag{3.41}$$

where $\tilde{\mathbf{X}}$ is the $n \times N$ rectangular matrix of normalized indicator vectors similar to (3.39),

$$\tilde{\mathbf{X}} = \left[\frac{\chi_1}{\sqrt{\mu(\Gamma_1)}}, \dots, \frac{\chi_n}{\sqrt{|\mu(\Gamma_n)|}}\right]. \tag{3.42}$$

The Weighted Ratio Association is reduced to the trace maximization of

$$\text{WRAssoc}(G) = \max_{\tilde{\mathbf{X}}^{\top}\tilde{\mathbf{X}}=\mathbf{1}_n} \text{tr}(\tilde{\mathbf{X}}^{\top}\mathbf{D}^{-1/2}\mathbf{L}\mathbf{D}^{-1/2}\tilde{\mathbf{X}}). \tag{3.43}$$

The optimization of graph partitioning objectives gives us graph cuts balanced with respect to the number of nodes in the subgraph Γ_i, their cumulative weights $|\Gamma_i|$, or their sizes $\mu(\Gamma_i)$ with respect to a certain measure assigned to nodes.

In the Chap. 2, we discussed that random walks defined on a connected undirected graph G set up the structure of $(N-1)$–dimensional Euclidean space such that for every pair of nodes $i \neq j$ we can introduce the positive symmetric distance $K(i,j) > 0$ (the commute time) (2.75). The probabilistic distance associated to random walks and the probabilistic angle (2.72) between them can be used as measures of similarity between two nodes in the graph G for the purpose of graph partitioning.

In the probabilistic space associated with random walks, each node of the graph G is represented by a certain vector $\mathbf{z}_i \in \mathbb{R}^{N-1}$. Then, we can assign each vector $\mathbf{z}_i$ to the cluster Γ_i whose center,

$$\mathbf{m}_i = \sum_{s=1}^{|\Gamma_i|} \frac{z_{is}}{|\Gamma_i|}, \tag{3.44}$$

(also called centroid) is nearest with respect to the graph random walks distance (2.75).

The objective we try to achieve is to minimize the total intra-cluster variance of the resulting partition $\mathfrak{P}$ of the graph G into n clusters, or the squared error function (s.e.f.),

$$\text{sef}(\mathfrak{P}) = \sum_{i=1}^{n} \sum_{s=1}^{|\Gamma_i|} |z_{is} - \mathbf{m}_i|^2. \tag{3.45}$$

Let $\mathbf{e} = (1, 1, \dots, 1)^\top$ be a $(N-1)$-dimensional vector of ones, and $\mathbf{Z} = (\mathbf{z}_i)$ be the $(N-1) \times N$ matrix of node coordinates. Then it is clear that

$$\begin{aligned} \text{sef}(\mathfrak{P}) &= \sum_{i=1}^{n} |\mathbf{Z}_i - \mathbf{m}_i \mathbf{e}^\top|^2 \\ &= \sum_{i=1}^{n} |\mathbf{Z}_i \mathbf{P}_i|^2, \end{aligned} \tag{3.46}$$

where

$$\mathbf{P}_i = \left(\mathbf{1}_{\Gamma_i} - \frac{\mathbf{e} \mathbf{e}^\top}{|\Gamma_i|} \right),$$

is the projection operator onto the cluster Γ_i. Since $\mathbf{P}_i^2 = \mathbf{P}_i$, we obtain

$$\begin{aligned} \text{sef}(\mathfrak{P}) &= \sum_{i=1}^{n} \text{tr}(\mathbf{Z}_i^\top \mathbf{P}_i \mathbf{Z}_i) \\ &= \text{tr}(\mathbf{Z}^\top \mathbf{Z}) - \text{trace}(\mathbf{X}^\top \mathbf{Z}^\top \mathbf{Z} \mathbf{X}), \end{aligned} \tag{3.47}$$

in which $\mathbf{X}$ is the $(N-1) \times N$ orthogonal indicator matrix (3.39).

If we consider the elements of the $\mathbf{Z}^\top \mathbf{Z}$ matrix as measuring similarity between nodes, then it can be shown (Zha et al. 2001) that Euclidean distance leads to Euclidean inner-product similarity which can be replaced by a general Mercer kernel (Saitoh 1988, Wahba 1990) uniquely represented by a positive semi-definite matrix $K(i, j)$. If we relax the discrete structure of $\mathbf{X}$ and assume that $\mathbf{X}$ is an arbitrary orthonormal matrix, the minimization of sef($\mathfrak{P}$) function is reduced to the trace maximization problem,

$$\max_{\mathbf{X}^\top \mathbf{X} = \mathbf{1}_{N-1}} \text{tr}(\mathbf{X}^\top \mathbf{Z}^\top \mathbf{Z} \mathbf{X}), \tag{3.48}$$

as it was the case for the graph partitioning according to the minimization of objective functions (3.33), (3.34), (3.35) and (3.36).

A standard result in linear algebra (proven by Fan 1949) provides a global solution to a related version of the trace optimization problems: Given a symmetric matrix $\mathbf{S}$ with eigenvalues $\lambda_1 \geq \ldots \geq \lambda_n \geq \ldots \geq \lambda_N$, and the matrix of corresponding eigenvectors, $[\mathbf{u}_1, \ldots, \mathbf{u}_N]$, the maximum of $\mathrm{tr}(\mathbf{Q}^\top \mathbf{S} \mathbf{Q})$ over all n-dimensional orthonormal matrices $\mathbf{Q}$ such that $\mathbf{Q}^\top \mathbf{Q} = \mathbf{1}_n$ is given by

$$\max_{\mathbf{Q}^\top \mathbf{Q} = \mathbf{1}_n} \mathrm{tr}(\mathbf{Q}^\top \mathbf{S} \mathbf{Q}) = \sum_{k=1}^{n} \lambda_k, \qquad (3.49)$$

and the optimal n-dimensional orthonormal matrix

$$\mathbf{Q} = [\mathbf{u}_1, \ldots, \mathbf{u}_n]\mathbf{R} \qquad (3.50)$$

where $\mathbf{R}$ is an arbitrary orthogonal $n \times n$ matrix (describing a rotation in $\mathbb{R}^n$).

The result (3.49), (3.50) relates the problem of network segmentations to the investigation of n primary eigenvectors of a symmetric matrix defined on the graph nodes (Golub et al. 1996, Dhillon et al. 2004b). The eigenvectors $\mathbf{u}_{i>1}$ have both positive and negative entries, so that in general $[\mathbf{u}_1, \ldots, \mathbf{u}_n]$ differ substantially from the matrix of discrete cluster indicator vectors which have strictly positive entries.

Even for not very large n it may be rather difficult to compute the appropriate $n \times n$ orthonormal transformation matrix $\mathbf{R}$ which would recover the necessary discrete cluster indicator structure. Furthermore, it can be shown that the post-processing of eigenvectors into the cluster indicator vectors can be reduced to an optimization problem with $n(n-1)/2 - 1$ parameters (Ding et al.). Several methods have been proposed to obtain the partitions from the eigenvectors of various similarity matrices (see Dhillon et al. 2004a and Bach et al. 2004 for a review).

In the next subsection, we use the ideas of Principal Component Analysis (PCA) in order to bypass the orthonormal transformation.

3.3 Principal Component Analysis of Venetian Canals

Spectral methods can be implemented in order to visualize graphs of not very large multicomponent networks (Volchenkov et al. 2007a). City districts constructed according to different development principles in different historical epochs can be envisioned by their spatial graphs.

3.3.1 Sestieri of Venice

We will discuss the segmentation of the canal network in Venice in accordance to its historical divisions. The sestieri are the primary traditional divisions of Venice (see Fig. 3.9): Cannaregio, San Polo, Dorsoduro, Santa Croce, San Marco and Castello, Giudecca. The oldest settlements in Venice had appeared from the 6th Century in Dorsoduro, along the Giudecca Canal. By the 11th Century, settlement had spread

Fig. 3.9 The sestieri are the primary traditional divisions of Venice. The image has been taken from 'Portale di Venezia' at *http://www.guestinvenice.com/*

to the Grand Canal. The Giudecca island is composed of eight islets separated by canals dredged in the 9th Century when the area was divided among the rebelling nobles. San Polo is the smallest of the six sestieri of Venice, covering just 35 hectares along the Grand Canal. It is one of the oldest parts of the city, having been settled before the 9th Century, when it and San Marco (lying in the heart of the city) formed part of the Realtine Islands. Cannaregio, named after the Cannaregio Canal, is the second largest district of the city. It was developed in the 11th Century. Santa Croce occupies the northwest part of the main islands lying on land created from the late Middle Ages to the 20th Century. The district Castello originated in the 13th Century.

While analyzing the canal network of Venice, we have assigned an identification number to each canal. Then the dual graph representation for the canal network is constructed by mapping canals encoded with the same ID into nodes of the spatial graph and intersections among each pair of canals in the city map into edges connecting the corresponding nodes of the spatial graph. We suppose for simplicity that the spatial graph is not weighted. Its segmentation is closely related to the problem of three-dimensional (3-D) visual representations.

In order to obtain the 3-D visual representation of the spatial graph for the canal network of Venice, we introduce the normalized Laplace operator (3.8) on the spatial graph and compute its eigenvectors $\{u_k\}$, $k > 1$, belonging to several of the smallest eigenvalues $\lambda_k > 0$. The (x_i, y_i, z_i) coordinates of the ith vertex of the spatial graph in 3-D space are given by the relevant ith components of three eigenvectors taken from $\{u_k\}$. Possible segmentations and symmetries of graphs can be discovered visually by using different triples of eigenvectors if the number of nodes in the graph is not too large. In Fig. 3.10, we have presented the results of segmentation for the spatial graph of Venetian canals using the eigenvectors $[u_2, u_3, u_4]$ belonging to the three smallest nontrivial eigenvalues $\lambda_2, \lambda_3, \lambda_4$. Nodes of the spatial graph (corresponding to canals on the primary map) belonging to one and the same city district, developed in a certain historical epoch, are located on one and the same quasi-surface in

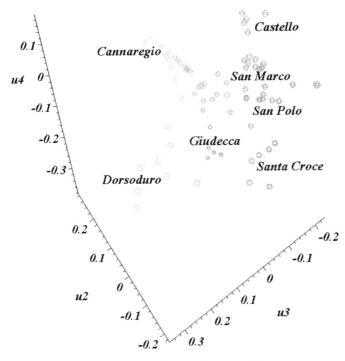

Fig. 3.10 The segmentation of the spatial graph of Venetian canals using three eigenvectors $[u_2, u_3, u_4]$. The nodes of the spatial graph can be partitioned into classes which can be almost precisely identified with the historical city districts

the multi-dimensional space spanned with the basis of eigenvectors of the Laplace operator. First eigenvectors of the normalized Laplace operator indicate the directions in which the transport network has maximal "extensions." The use of these eigenvectors as a basis helps us to divide the nodes of the spatial graph into classes which can be almost precisely identified with the historical city districts. Let us note that the implementation of other eigenvectors belonging to the eigenvalues λ_k, with $k \gg 1$, as the basis for the 3-D representation of the graph worsens the quality of segmentation.

The slowest modes of diffusion process allow one to detect the city modules characterized by the individual accessibility properties. The primary feature of the diffusion process in the compact urban patterns is the flow between the dominant pair of city modules: the "broadways" and the relatively isolated streets remote from the primary roads.

Due to proper normalization, the components of eigenvectors play the role of the Participation Ratios (PR) which quantify the effective numbers of nodes participating in a given eigenvector with a significant weight. This characteristic has been used in (Eriksen et al. 2003) and by other authors to describe the modularity of complex networks. However, PR is not a well-defined quantity in the case of eigenvalue multiplicity since the different vectors in the eigenspace corresponding to the degenerate mode would obviously have different PR.

3.3.2 A Time Scale Argument for the Number of Essential Vectors

Visual segmentation of transport networks based on the 3-D representation of their spatial graphs is not always feasible. Furthermore, the result of such a segmentation would essentially depend on which eigenvectors have been chosen as the basis for the 3-D representation. The computation of eigenvectors for large matrices can be time and resource consuming and, therefore, it is important to have a good estimation on the minimal number of eigenvectors required for the proper graph segmentation.

In statistics, Principal Component Analysis (PCA) is used to reduce the size of a data set. It is achieved by the optimal linear transformation retaining the subspace that has the largest variance (a lower-order principal component) and ignoring higher-order ones (Fukunaga 1990, Jolliffe 2002).

Given an operator S self-adjoint with respect to the measure m defined on a connected undirected graph G, it is well-known that the eigenvectors of the symmetric matrix $\mathbf{S}$ form an ordered orthonormal basis $\{\phi_k\}$ with real eigenvalues $\mu_1 \geq \ldots \geq \mu_N$. The ordered orthogonal basis represents the directions of the variances of variables described by S.

If we consider the Laplace operator, $L = 1 - T$, defined on G, its eigenvalues can be interpreted as the inverse characteristic time scales of the diffusion process such that the smallest eigenvalues correspond to the stationary distribution together with the slowest diffusion modes involving the most significant amounts of flowing commodity. Therefore, while describing a network by means of the Laplace operator, we must arrange the eigenvalues in increasing order, $\lambda_1 < \lambda_2 \leq \ldots \leq \lambda_n \leq \ldots \leq \lambda_N$, and examine the ordered orthogonal basis of eigenvectors, $[\mathbf{f}_1, \ldots \mathbf{f}_N]$.

The number of components which may be detected in a network with regard to a certain defined dynamical process depends upon the number of essential eigenvectors of the relevant self-adjoint operator. There is a simple time scale argument which we use in order to determine the number of applicable eigenvectors.

It is obvious that while observing the network close to an equilibrium state for a short time, we detect flows resulting from a large number of transient processes evolving toward the stationary distribution and being characterized by the relaxation times $\propto 1/\log \lambda_k$. While measuring the flows over a insufficiently long time τ, we may discover just n different eigenmodes, such that

$$\lambda_1 < \ldots \leq \lambda_n \leq \exp\left(-\tau^{-1}\right) < \ldots \leq \lambda_N. \qquad (3.51)$$

In general, the longer the time of measurements τ, the less is the number of eigenvectors we must to take into account in component analysis of the network. If the time of measurements is fixed, we can determine the number of required eigenvectors.

3.3.3 Low-Dimensional Representations of Transport Networks by Principal Directions

In order to obtain the best quality segmentation, it is convenient to center the n primary eigenvectors. The centroid vector (representing the center of mass of the set $[\mathbf{f}_1, \ldots \mathbf{f}_n]$) is calculated as the arithmetic mean,

$$\mathbf{m} = \frac{1}{n} \sum_{k=1}^{n} \mathbf{f}_k. \tag{3.52}$$

Let us denote the $n \times N$ matrix of n centered eigenvectors by

$$\mathbf{F} = [\mathbf{f}_1 - \mathbf{m}, \ldots \mathbf{f}_N - \mathbf{m}].$$

Then, the symmetric matrix of covariances between the entries of eigenvectors $\{\mathbf{f}_k\}$ is the product of $\mathbf{F}$ and its adjoint $\mathbf{F}^\top$,

$$\mathbf{Cov} = \frac{\mathbf{F}\mathbf{F}^\top}{N-1}. \tag{3.53}$$

The correspondent Gram matrix

$$\mathbf{F}^\top \mathbf{F}/(N-1) \equiv \mathbf{1}$$

due to the orthogonality of the basis eigenvectors. The main contributions in the symmetric matrix $\mathbf{Cov}$ are related to the groups of nodes

$$\mathbf{Cov}\,\mathbf{u}_k = \sigma_k\,\mathbf{u}_k, \tag{3.54}$$

which can be identified by means of the eigenvectors $\{\mathbf{u}_k\}$ associated to the first largest eigenvalues among $\sigma_1 \geq \sigma_2, \ldots, \geq \sigma_N$. By ordering the eigenvectors in decreasing order (largest first), we create an ordered orthogonal basis with the first eigenvector having the direction of the largest variance of the components of n eigenvectors $\{\mathbf{f}_k\}$. Note that, due to the structure of $\mathbf{F}$, only the first $n-1$ eigenvalues σ_k are not trivial. In accordance with the standard PCA notation, the eigenvectors of the covariance matrix $\mathbf{u}_k$ are called the *principal directions* of the network with respect to the diffusion process defined by the operator S. A low-dimensional representation of the network is given by its principal directions $[\mathbf{u}_1, \ldots, \mathbf{u}_{n-1}]$, for $n < N$.

Diagonal elements of the matrix $\mathbf{Cov}$ quantify the component variances of the eigenvectors $[\mathbf{f}_1, \ldots \mathbf{f}_n]$ around their mean values (3.52) and may be ample essentially for large networks. Therefore, it is practical for us to use the standardized correlation matrix,

$$Corr_{ij} = \frac{Cov_{ij}}{\sqrt{Cov_{ii}}\sqrt{Cov_{jj}}}, \tag{3.55}$$

instead of the covariance matrix **Cov**. The diagonal elements of (3.55) equal 1, while the off-diagonal elements are the Pearson's coefficients of linear correlations (Cohen et al. 2003).

The correlation matrix (3.55) calculated with regard to the first n eigenvectors possesses a complicated structure containing the multiple overlapping blocks pertinent to a low-dimensional representation of the network of Venetian canals which allows for a further simplification. In Fig. 3.11, we have presented the correlation matrix (3.55) calculated for the first seven eigenvectors of the normalized Laplace operator (3.8) defined on the dual graph representation of 96 Venetian canals.

Let **U** be the orthonormal matrix which contains the eigenvectors $\{\mathbf{u}_k\}$, $k = 1,\ldots, n-1$, of the covariance (or correlation) matrix as the row vectors. These vectors form the orthogonal basis of the $(n-1)$-dimensional vector space, in which every variance $(\mathbf{f}_k - \mathbf{m})$ is represented by a point $\mathbf{g}_k \in \mathbb{R}^{(n-1)}$,

$$\mathbf{g}_k = \mathbf{U}(\mathbf{f}_k - \mathbf{m}). \tag{3.56}$$

Then each original eigenvector $\mathbf{f}_k$ can be decoded from $\mathbf{g}_k \in \mathbb{R}^{(n-1)}$ by the inverse transformation,

$$\mathbf{f}_k = \mathbf{U}^\top \mathbf{g}_k + \mathbf{m}. \tag{3.57}$$

The use of transformations (3.56) and (3.57) allows us to obtain the $(n-1)$-dimensional representation $\{\varphi_k\}_{k=1}^{(n-1)}$ of the N-dimensional basis vectors $\{\mathbf{f}_s\}_{s=1}^{N}$ in the form

$$\varphi_k = \mathbf{U}^\top \mathbf{U}\mathbf{f}_k + \left(\mathbf{1} - \mathbf{U}^\top \mathbf{U}\right)\mathbf{m}, \tag{3.58}$$

that minimizes the mean-square error between $\mathbf{f}_k \in \mathbb{R}^N$ and $\varphi_k \in \mathbb{R}^{(n-1)}$ for given n.

Variances of eigenvectors $\{\mathbf{f}_k\}$ are positively correlated within a principal component of the transport network. Thus, the transition matrix $\mathbf{U}^\top \mathbf{U}$ can be interpreted as the connectivity patterns acquired by the network with respect to the diffusion process. Two nodes, i and j, belong to one and the same principal component of the

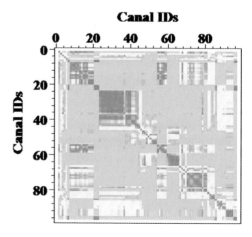

Fig. 3.11 The correlation matrix calculated for the spatial graph of Venetian canals calculated for the system of the first seven eigenvectors of the normalized Laplace operator (3.8). The entries of the matrix are ranked from 1 (red) to -1 (cyan)

Fig. 3.12 The coarse-grained connectivity matrix derived from the low-dimensional representation of Venetian canals given by the transition matrix $\mathbf{U}_6\,\mathbf{U}_6^\top$

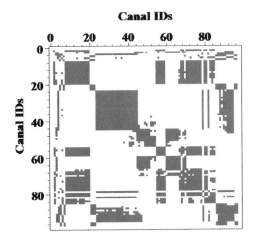

network if $\left(\mathbf{U}\mathbf{U}^\top\right)_{ij} > 0$. By applying the Heaviside function–which is zero for negative argument–and one for positive argument, to the elements of the transition matrix $\mathbf{U}\mathbf{U}^\top$, we derive the coarse-grained connectivity matrix of network components. In Fig. 3.12, we have shown the coarse-grained connectivity matrix obtained from the transition matrix $\mathbf{U}_6\,\mathbf{U}_6^\top$ for the dual graph representation of Venetian canals.

3.3.4 Dynamical Segmentation of Venetian Canals

In general, the building of low-dimensional representations for transport networks with respect to a certain dynamical process defined on them is a complicated procedure which cannot be reduced to (and reproduced by) the naive introduction of "supernodes" by either merging of several nodes or shrinking complete subgraphs of the original graph.

If the covariance matrix clearly exhibits a block structure, and once the relevant coarse-grained connectivity matrix is computed, we can identify dynamical clusters (blocks) by using a linearized cluster assignment and compute the cluster crossing, the cluster overlap along the specified ordering using the spectral ordering algorithm (Ding et al. 2004). The problem of dynamical segmentations of a transport network in fast time scales is more computationally complex, especially for large networks, because many eigenvectors, if not all, have to be taken into account while calculating the covariance matrix. It is important to note that the covariance matrix in this case takes the form of a sparse, nearly diagonal matrix (see Fig. 3.13).

Sparsity of the deduced coarse-grained connectivity matrix (which is shown in Fig. 3.14) in fast time scales entails loosely coupled systems that lack any form of large scale structure. A sparse coarse-grained connectivity matrix may be useful when storing and manipulating data for approximate descriptions of transport networks in fast time scales.

Fig. 3.13 The covariance
matrix calculated for the
spatial graph representation
of Venetian canals with
regrades to the first 80
eigenvectors of the
normalized Laplace operator
(3.8). The entries of
covariance matrix are ranked
from the largest positive
values (red) to the utmost
negative values (*blue*)

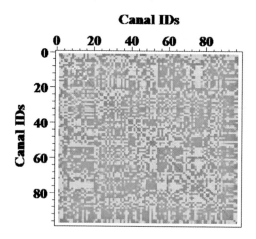

Low-dimensional representations of not very large transport networks given by
the coarse-grained connectivity matrices can be envisioned by the 3-D graphs.

In Fig. 3.15, we have shown the 3-D figure representing the dynamical segmentation of Venetian canals by blobs. Each blob corresponds to a canal on the city plan
of Venice. The radius of the blob has been taken equal to the first-passage time to
the canal by random walks on the spatial graph of the city canal network. The nodes
characterized by the worst accessibility property have the longest first-passage times
are, therefore, represented by the largest balls. Consequently, the canals distinguished by the shortest first-passage times are represented by the smallest blobs,
which are not seen in Fig. 3.15 being immersed into the larger blobs.

The components of the three major eigenvectors of the coarse-grained connectivity matrix shown in Fig. 3.14 have been used as three Cartesian coordinates of
each blob in Fig. 3.15. These eigenvectors determine the direction of the largest
variances of correlations delineating the low-dimensional representation of the network. Although the canals characterized by the relatively short access times are also

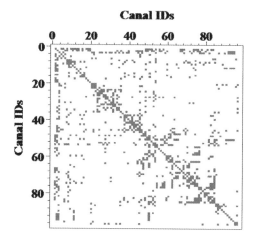

Fig. 3.14 The fast time scale
coarse-grained connectivity
matrix for the
low-dimensional
representation of Venetian
canals deduced from the
transition matrix $\mathbf{U}_{79}\mathbf{U}_{79}^{\top}$

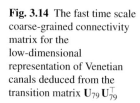

Fig. 3.15 The 3-D image of a dynamical segmentation of Venetian canals built for the first eight eigenvalues of the normalized Laplace operator (3.8). The structural differences between the historical city districts of Venice are clearly visible

characterized by small variances of correlations, they are forgathered proximate to the center of the figure displayed in Fig. 3.15, no matter which city district they belong to. In contrast, the worst accessible canals are distinguished by the strongest correlation variances and are located on the figure fringes, far apart from its center. At the same time, the radiuses of these blobs are the largest among all other blobs since they acquire the utmost norms with respect to random walks.

It is remarkable that the trends of variances clearly visible in Fig. 3.15 can be perfectly identified with the traditional historical sestieri of Venice.

3.4 Thermodynamical Formalism for Urban Area Networks

In this section, we study the spectra

$$\rho(\lambda) = \sum_{k=1}^{N} \delta(\lambda - \lambda_k), \quad \lambda_k \in [0,2], \tag{3.59}$$

of the normalized Laplace operator (3.8) defined on the connected, undirected spatial graphs of compact urban patterns.

3.4.1 In Search of Lost Time: Is there an Alternative for Zoning?

The profound motivation of our study is to develop a method for the multiple time scale analysis of urban fabric.

The urban fabric is the physical form of towns and cities within which many modes of transportation are accommodated. Each mode can be characterized by its specific time scale.

In walkable communities, acting on the primary, "slowest" time scale, the urban fabric should encourage walking as the primary mode of transportation. However, this only seems to be possible when local services are primarily accessed by walking, and the walking environment is safe and pleasant for pedestrians.

It is nevertheless clear that walkable communities are only economically feasible when an infrastructure of quality transit services is provided. Therefore, it is inevitable that cars and cargo vehicles populating another, "fast" time scale transportation mode should also be accommodated in most walkable communities, but not to the extent that they jeopardize the goals of walkability.

The traditional planning strategy for achieving these accommodation goals is zoning, or the separation of land uses into different areas achieved by limiting the amount of land used for roads and parking, and by building multistory buildings. In particular, this tendency would lead to the excessive density of urban fabric, such as that found in Manhattan.

When distances are small they are likely to be covered on foot. This is indeed necessary for certain uses, such as heavy industry, but the effect of separating all other uses from one another is that one can no longer walk from one use to another.

By allowing a diverse range of uses in the same areas, walking distances may be further reduced, and transit can be made more viable because the same transportation lines can be used by different people and different traffic flows at different times of day.

In the present section, we use the thermodynamic formalism in order to describe the possible effect of urban texture on a complex transportation process at many time scales at once.

The obvious advantage of statistical mechanics is that statistical moments of random walks would acquire a "physical interpretation" in the framework of the thermodynamic formalism. The full description of transport networks in short time and small scales requires a high dimensional space, i.e., the knowledge of locations and velocities of all agents participating in transport processes. However, if being defined on a graph, the time evolution of such a system can be described by just a few dynamically relevant variables called reaction coordinates (Nadler et al. 2006).

Identifying slow variables and dynamically meaningful reaction coordinates that capture the long-term evolution of transport systems is among the most important problems of transport network analysis. The spectrum of the Laplace operator could have gaps indicating the time scale's separation, that is, there are only a few "slow" time scales at which the transport network is metastable, with many "fast" modes describing the transient processes toward the slow modes. The methods of spectral graph theory allow detecting and separating "slow" and "fast" time scale's giving rise to the component analysis of networks in which the primary eigenvalues play the essential role.

The spectrum (3.59) is nonnegative and, therefore, we can investigate it by means of methods strongly inspired by statistical mechanics. The characteristic function,

$$\begin{aligned} Z(\beta) &= \int_0^\infty e^{-\beta\lambda} \rho(\lambda)\, d\lambda \\ &= \sum_{k=1}^N e^{-\beta\lambda_k}, \end{aligned} \tag{3.60}$$

(known as the canonical partition function) discriminates large eigenvalues in favor of smaller ones. The laziness parameter β determines the time scale in the random walk process (the mobility of random walkers).

For random walks defined on the undirected graphs, the density distributions of random walkers, $\sigma_i \geq 0$, $\sum_{i \in V} \sigma_i = 1$, play the role of the microscopic observables. A system is in equilibrium when its macroscopic observables have ceased to change with time, thus the stationary distribution of random walks $\pi = \psi_1^2$ can be naturally interpreted as the equilibrium state of the random walk process. Other microstates are described by the next eigenvectors.

The stationary distribution of random walks is the state of maximal probability and, therefore, bringing up the concept of the canonical ensemble, it is possible to derive the probability Pr_k that random walkers could be found on the graph G in a certain microstate $\sigma_i^{(k)} = \psi_{k,i}^2$ belonging to the spectral value λ_k:

$$\mathrm{Pr}_k = Z^{-1}(\beta)\exp(-\beta\lambda_k). \tag{3.61}$$

Since random walkers neither have masses, nor kinetic energy, they do not interact with each other and therefore non trivial expectations which can be found with the use of (3.60) characterize nothing else but certain structural properties of the graph itself.

3.4.2 Internal Energy of Urban Space

The averaged eigenvalue,

$$\begin{aligned}
\langle \lambda \rangle &= Z(\beta)^{-1} \sum_{i=1}^{N} \lambda_i e^{-\beta\lambda_i} \\
&= -\partial_\beta \ln Z(\beta),
\end{aligned} \tag{3.62}$$

held at a constant value of laziness parameter β can be interpreted as the microscopic definition of the thermodynamic variable corresponding to the internal energy in statistical mechanics.

Due to the complicated topology of streets and canals, the flow of random walkers exhibits spectral properties similar to that of a thermodynamic system characterized by a nontrivial internal energy.

The principle of equipartition of energy in classical statistical mechanics gives the average values of individual components of the energy, such as the kinetic energy of a particular particle- it can be applied to any classical system in thermal equilibrium, no matter how complicated. In particular, the equipartition theorem states that each molecular quadratic degree of freedom receives $1/2\,kT = 1/(2\beta)$ of energy.

From Fig. 3.16, one can clearly see that at low mobility of random walkers ($\beta \ll 1$) the system of lazy random walks defined on the spatial graphs of compact urban patterns behaves as a system characterized by two quadratic degrees of freedom. While β increases, more random walkers probably change their locations at each time step, thus contributing a pattern of motion characterized

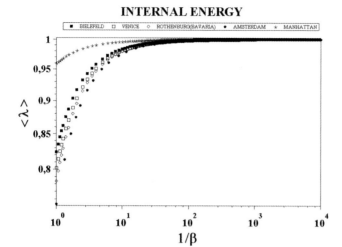

Fig. 3.16 The expected values of the "microscopic energy" (the averaged eigenvalues) are calculated for the spectra of compact urban patterns via the laziness of random walkers

by one degree of freedom. The difference between the street layout patterns of the organic cities (Bielefeld, Rothenburg, canals of Amsterdam and Venice) and the street grid of Manhattan is revealed by the relative decrease of the internal energy as $\beta \to 1$. While in Manhattan, random walks remain almost two-dimensional even at $\beta = 1$.

Taking the derivative of $\langle \lambda \rangle$ with respect to the parameter β in (3.62), we have

$$\frac{d\langle \lambda \rangle}{d\beta} = \langle \lambda \rangle^2 - \langle \lambda^2 \rangle = -D^2(\lambda), \tag{3.63}$$

where $D^2(\lambda)$ is the variance, the measure of its statistical dispersion, indicating how the eigenvalues of the normalized Laplace operator are spread around the expected value $\langle \lambda \rangle$ tracing the variability of the eigenvalues. The standard deviation is then calculated by

$$\sigma = \sqrt{D^2(\lambda)}. \tag{3.64}$$

A large standard deviation indicates that the data points are far from the mean and a small standard deviation indicates that they are clustered closely around the mean (see Fig. 3.17).

The standard deviations of eigenvalues around the mean are almost insensitive to the mobility of random walkers and are twice as larger for organic cities than for the pattern planned in a regular grid (see Fig. 3.17).

3.4.3 Entropy of Urban Space

In thermodynamics, entropy accounts for the effects of irreversibility describing the number of possible observables in the system. For random walks defined on finite undirected graphs, its value,

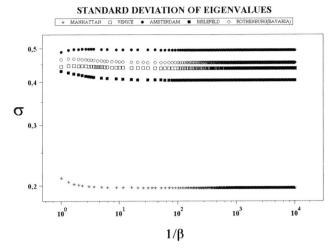

Fig. 3.17 The standard deviations of eigenvalues around the mean are calculated for the spectra of compact urban patterns via the laziness of random walkers

$$S = -\sum_{k=1}^{N} \Pr_k \ln \Pr_k, \qquad (3.65)$$

quantifies the probable number of density distributions of random walks which can be observed in the system before the stationary distribution π is achieved.

In Fig. 3.18, we have displayed the entropy curves versus inverse mobility in the studied cities.

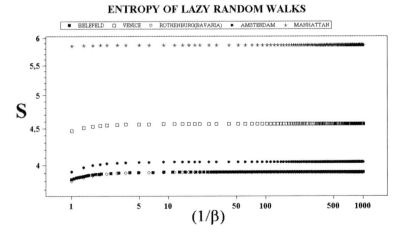

Fig. 3.18 The entropy of lazy random walks calculated for the spectra of compact urban patterns via the laziness of random walkers

3.4.4 Pressure in Urban Space

Due to the numerous junctions and the highly entangled meshes of streets and canals, random walkers "loss their ways," so that it takes them longer to cross the city while roaming randomly along the routes. This time delay can be interpreted as an effect of a slight negative pressure involving the random walkers into the city. It is obvious that the strength of pressure should vary from one district to another within a city and can be different for different eigenmodes of the diffusion process with no rigorous bind to the city administrative units being localized on certain groups of routes.

The value of pressure is positive if free energy decreases with the network size and negative otherwise. Pressure can be calculated by

$$\mathscr{P} = -\left(\frac{\partial F}{\partial N}\right) = \frac{1}{\beta}\left(\frac{\partial \ln Z(\beta)}{\partial N}\right) \tag{3.66}$$

where N is the network size. The spectral density function $\rho(\lambda)$ calculated for the urban textures formally relates the extensive thermodynamic quantities with their spectral analogs,

$$dN \simeq \rho(\lambda)\,d\lambda. \tag{3.67}$$

The use of continuous approximations for the spectral density function allows for computation of the pressure spectra $P(\lambda)$ forcing the flows of random walkers with eigenmodes λ into the compact city structures (see Figs. 3.19, 3.20 and 3.21).

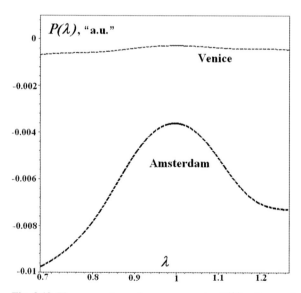

Fig. 3.19 The comparison of pressure spectra $P(\lambda)$ acting on the flows of random walkers with eigenmodes λ into Amsterdam and Venice.

Fig. 3.20 The comparison of pressure spectra $P(\lambda)$ acting on the flows of random walkers with eigenmodes λ into Bielefeld and Rothenburg

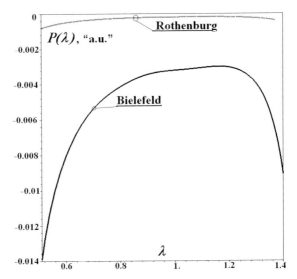

Generally speaking, the more junctions a city has, the stronger the drag force (pressure): even though venices city network has more canals than in Amsterdam, the number of junctions between them is less than in Amsterdam and, therefore, the negative pressure in the latter network is stronger (Fig. 3.19). The numbers of streets in Rothenburg and in the downtown of Bielefeld are equal, but there are more crossroads in Bielefeld than in Rothenburg (Fig. 3.20).

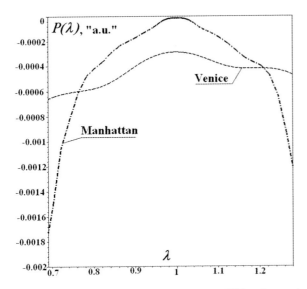

Fig. 3.21 The comparison of pressure spectra $P(\lambda)$ acting on the flows of random walkers with eigenmodes λ into Manhattan and Venice

One can see that pressure profiles have maxima close to $\lambda = 1$ (correspondent to a minimal drug force) for Amsterdam, Venice, and Manhattan. It calls for the idea of a "transparency corridor," i.e., a sequence of streets and junctions (on which the relevant eigenmodes are localized) along which the city can be crossed within minimal time. In Manhattan, the pressure profile is almost zero with $\lambda = 1$. The multiple eigenvalue $\lambda = 1$ indicates the segments of high regularity in the urban texture which are indistinguishable with respect to the random walker dynamics. In Venice, the minimal pressure is achieved on the Grand Canal and Giudecca Canal, and it is due to Het IJ and Amstel river in Amsterdam.

3.5 Summary

We have used the spectral characteristics of Laplace operators for the purpose of community detection in the spatial graphs of urban environments.

Eigenvalues of the normalized Laplace operator are real and bounded in the interval $[0, 2]$, and their distributions are dissimilar to the spectra observed for random and real-world graphs studied thus far.

The probability distributions of eigenvalues in the spectra of organic cities are akin to the Gaussian curve centered at 1. The spectra of these city canal networks look amazingly alike even though they were founded in dissimilar geographical regions and for different purposes.

We have also estimated the Cheeger constant, a numerical measure of whether or not a graph has a "bottleneck," for the compact urban patterns and compared them with Cheeger's constants estimated for connected realizations of the random graphs $G(N, M)$.

Spectral methods can be implemented in order to visualize graphs multicomponent networks that are not very large.

Chapter 4
Spectral Analysis of Directed Graphs and Interacting Networks

In the preceding chapters, we have discussed networks abstracted as undirected graphs. However, many transport networks naturally correspond to directed graphs. Power grids transporting electrical currents and driving directions assigned to city streets in order to optimize traffic in the city are the examples (Albert et al. 2002).

Traffic within large cities is formally organized with marked driving directions creating one-way streets. It is well-known empirically that the use of one-way streets would greatly improve traffic flow since the speed of traffic is increased and intersections are simplified.

Moreover, even if a network of city itineraries formally constitutes an undirected graph, there are the traffic negotiation principles such as "first come, first served" and "yielding to crossing traffic" that establish goes first while crossing the road intersections or other conflicting parts of the road and who should wait until the other driver does so. These yielding rules that are common knowledge among drivers and are intended to prevent collisions both between intersections and at intersections perplex the instantaneous traffic analysis by converting the network of city itineraries into a complex transport system with dynamically altering directions.

It is also important to mention that the traffic negotiation principles differ between countries and, therefore, the actual traffic analysis should always take regional features into account. For instance, about 34 percent of the world's population drives on the left, and 66 percent on the right. By roadway distances, about 28 percent drive on the left, and 72 percent on the right (see Lucas 2005). While introducing weights estimating traffic loading of itineraries in the city spatial graph, we soon discover that each road side is characterized by its own weight depending upon how many right-hand and left-hand junctions the road side has with other itineraries, so that an honest representation of any transport network is usually given by a weighted directed graph.

To our knowledge, there are only few works devoted to the analysis of complex directed networks. Furthermore, the global transport network in the city usually consists of many different transportation modes which can be used alternatively by passengers while travelling (say, by tram, bus, subway, private vehicle, and by foot) at certain transport stations or elsewhere. Interactions between the individual transportation modes by means of passengers are responsible for the multiple complex phenomena we may observe each day on streets and while using the public transportation.

Ph. Blanchard, D. Volchenkov, *Mathematical Analysis of Urban Spatial Networks,*
Understanding Complex Systems, DOI 10.1007/978-3-540-87829-2_4,
© Springer-Verlag Berlin Heidelberg 2009

4.1 The Spectral Approach For Directed Graphs

The spectral approach for directed graphs has not been as well developed as undirected graphs because it is rather difficult, if ever possible, to define *unique* self-adjoint operator on directed graphs. In general, any node i in a directed graph $\overrightarrow{G}$ can have a different number of in-neighbors and out-neighbors,

$$\deg_{\text{in}}(i) \neq \deg_{\text{out}}(i). \tag{4.1}$$

In particular, a node i is a *source* if $\deg_{\text{in}}(i) = 0$, $\deg_{\text{out}}(i) \neq 0$, and is a *sink* if $\deg_{\text{out}}(i) = 0$, $\deg_{\text{in}}(i) \neq 0$. If the graph has neither sources nor sinks, it is called *strongly connected*. A directed graph is called Eulerian if $\deg_{\text{in}}(i) = \deg_{\text{out}}(i)$ for $\forall i \in \overrightarrow{G}$.

In general, the local structure of directed graphs is fundamentally different from that of undirected graphs. In particular, the diameters of directed networks can essentially exceed those of the same networks regarded as undirected. A recent investigation by Bianconi et al. (2007) shows that directed networks often have very few short loops, compared to finite random graph models.

In undirected networks, the high density of short loops (high clustering coefficient) together with a small graph diameter gives rise to the small-world effect (Watts et al. 1998). In directed networks, the correlation between the number of incoming and outgoing edges modulates the expected number of short loops. In particular, it has been demonstrated in Bianconi et al. (2007) that if the values $\deg_{\text{in}}(i)$ and $\deg_{\text{out}}(i)$ are not correlated, then the number of short loops, is strongly reduced, compared to when the degrees are positively correlated.

4.2 Random Walks on Directed Graphs

Finite random walks are defined on a strongly connected directed graph $\overrightarrow{G}(V, \overrightarrow{E})$ as finite vertex sequences $\mathfrak{w} = \{v_0, \ldots, v_n\}$ (time forward) and $\mathfrak{w}' = \{v_{-n}, \ldots, v_0\}$ (time backward) such that each pair (v_{i-1}, v_i) of vertices adjacent either in $\mathfrak{w}$ or in $\mathfrak{w}'$ constitutes a directed edge $v_{i-1} \to v_i$ in $\overrightarrow{G}$.

4.2.1 A Time–Forward Random Walk

A time–forward random walk is defined by the transition probability matrix (Chung 2005) P_{ij} for each pair of nodes $i, j \in \overrightarrow{G}$ by

$$P_{ij} = \begin{cases} 1/\deg_{\text{out}}(i), & i \to j, \\ 0, & \text{otherwise,} \end{cases} \tag{4.2}$$

which satisfies the probability conservation property:

$$\sum_{j,\ i\to j} P_{ij} = 1. \tag{4.3}$$

The definition (4.2) can be naturally extended for weighted graphs (Chung 2005) with $w_{ij} > 0$,

$$P_{ij} = \frac{w_{ij}}{\sum_k w_{ik}}. \tag{4.4}$$

Matrices (4.2) and (4.4) are real, but not symmetric and, therefore, have complex conjugated pairs of eigenvalues. For each pair of nodes $i, j \in \overrightarrow{G}$, the *forward* transition probability is given by $p_{ij}^{(t)} = (\mathbf{P}^t)_{ij}$ that is equal zero, if $\overrightarrow{G}$ contains no directed path from i to j.

4.2.2 Backwards Time Random Walks

Backwards time random walks are defined on the strongly connected directed graph $\overrightarrow{G}$ by the stochastic transition matrix

$$P_{ij}^{\star} = \begin{cases} 1/\deg_{in}(i), & j \to i, \\ 0, & \text{otherwise}, \end{cases} \tag{4.5}$$

satisfying *another* probability conservation property

$$\sum_{i,\ i\to j} P_{ij}^{\star} = 1. \tag{4.6}$$

It describes random walks unfolding backwards in time: should a random walker arrive at $t = 0$ at a node v_0, then

$$p_{ij}^{(-t)} = ((\mathbf{P}^{\star})^t)_{ij} \tag{4.7}$$

defines the probability that t steps *before* it had originated from a node j. The matrix element (4.7) is zero, provided there is no directed path from j to i in $\overrightarrow{G}$.

It is well–known that the evolution of densities $f \in \mathbb{R}^N$ such that $\sum_{v \in V} f(v) = 1$ in systems for which the dynamics are deterministic may be studied by the use of the linear Perron-Frobenius and Koopman operators (Mackey 1991). Strongly connected directed graphs $\overrightarrow{G}(V, \overrightarrow{E})$ can be interpreted as the discrete time dynamical systems specified by the dynamical law $\mathscr{S} : V \to V$,

$$\mathscr{S}(A) = \{w \in V \,|\, v \in A, v \to w\}. \tag{4.8}$$

Therefore, the transition operators of random walks defined above can be readily related to the Perron-Frobenius and Koopman operators (Koopman 1931) (the reference has been given in Mackey (1991). In particular, the Perron-Frobenius operator P^t, $t \in \mathbb{Z}_+$, transports the density function f, supported on the set A, forward in time to a function supported on some subset of $\mathscr{S}_t(A)$,

$$\sum_{v \in A} P^t f(v) = \sum_{\mathscr{S}_t^{-1}(A)} f(v).$$ (4.9)

The Koopman operator adjoint to the Perron-Frobenius one is thought of as transporting the density f supported on A backwards in time to a density supported on a subset of $\mathscr{S}_t^{-1}(A)$:

$$(P^\star)^t f(v) = f(\mathscr{S}_t(v)).$$ (4.10)

4.2.3 Stationary Distributions on Directed Graphs

If $\overrightarrow{G}$ is strongly connected and *aperiodic*, the random walk converges (Lovasz et al. 1995, Chung 2005, Bjorner et al. 1991, Bjorner et al. 1992) to the *only* stationary distribution π that is the Perron-Frobenius vector,

$$\pi \mathbf{P} = \pi.$$ (4.11)

If the graph $\overrightarrow{G}$ is periodic, then the transition probability matrix $\mathbf{P}$ can have more than one eigenvalue with absolute value 1 (Chung 1997).

The components of Perron's vector (4.11) can be normalized in such a way that $\sum_i \pi_i = 1$; moreover there is a bound for the ratio of their maximal and minimal values (Chung 2005),

$$\frac{\max_i \pi_i}{\min_i \pi_i} \le \max_i \deg_{\text{out}}(i)^{\text{diam}(\overrightarrow{G})},$$ (4.12)

where $\text{diam}(\overrightarrow{G})$ denotes the diameter of the strongly connected graph $\overrightarrow{G}$. In particular, it follows from (4.12) that the Perron vector π for random walks defined on a strongly connected directed graph can have coordinates with exponentially small values.

Any nonnegative function $F_{i \to j}$ defined on the set of graph edges E is called a circulation if it enjoys the Kirchhoff's law at each node $i \in \overrightarrow{G}$,

$$\sum_{s, \{s \to i\}} F_{s \to i} = \sum_{j, \{i \to j\}} F_{i \to j}.$$ (4.13)

For a directed graph $\overrightarrow{G}$, the Perron eigenvector π satisfies

$$F_{i \to j} = \pi_i P_{ij}$$ (4.14)

for any edge $i \to j$ (Chung 2005).

4.3 Laplace Operator Defined on the Aperiodic Strongly Connected Directed Graphs

Given an aperiodic strongly connected graph $\vec{G}$, a self-adjoint Laplace operator $L = L^{\top}$ can be defined (Chung 2005) as

$$L_{ij} = \delta_{ij} - \frac{1}{2}\left(\pi^{1/2}\mathbf{P}\pi^{-1/2} + \pi^{-1/2}\mathbf{P}^{\top}\pi^{1/2}\right)_{ij}, \tag{4.15}$$

where π is a diagonal matrix with entries π_i. The matrix (4.15) is symmetric and has real nonnegative eigenvalues $0 = \lambda_1 < \lambda_2 \leq \ldots \leq \lambda_N$ and real eigenvectors.

It can be easily proven (Butler 2007) that the Laplace operator (4.15) defined on the aperiodic strongly connected graph $\vec{G}$ is equivalent to the Laplace operator defined on a symmetric undirected weighted graph G_w on the same vertex set with weights defined by

$$w_{ij} = F_{i \to j} + F_{j \to i} \tag{4.16}$$
$$\equiv \pi_i P_{ij} + \pi_j P_{ji}.$$

Suppose that the transition probability matrix (4.2) has eigenvalues $\{\mu_i\}$, then it can be proven (Chung 2005) that

$$\min_{i \geq 2}(1 - |\mu_i|) \leq \lambda_2 \leq \min_{i \geq 2}(1 - \Re\mu_i), \tag{4.17}$$

where $\Re x$ denotes the real part of $x \in \mathbb{C}$.

Some results on spectral properties of Laplace operators can be translated to directed graphs. For instance, a Cheeger inequality for a directed graph has been established in Chung (2005). Given a subset of vertices $\Gamma \subset \vec{G}$, the out-boundary of Γ is

$$\partial\Gamma = \{i \to j, \, i \in \Gamma, j \in \vec{G} \setminus \Gamma\}. \tag{4.18}$$

The circulation through the out-boundary equals

$$F_{\partial\Gamma} = \sum_{i \in \Gamma, \, j \in \vec{G} \setminus \Gamma} F_{i \to j}, \tag{4.19}$$

and let

$$F_{\Gamma} \equiv \sum_{i \in \Gamma} F_i \equiv \sum_{j, \, j \to i} F_{j \to i}. \tag{4.20}$$

For a strongly connected graph $\vec{G}$ with stationary distribution π, the Cheeger constant can be defined as

$$h(\vec{G}) = \inf_{\Gamma \subset \vec{G}} \frac{F_{\partial\Gamma}}{\min_{\Gamma}(F_{\Gamma}, F_{\vec{G} \setminus \Gamma})}. \tag{4.21}$$

Given λ, the minimal nontrivial eigenvalue of Laplace operator (4.15), the Cheeger constant (4.21) satisfies the following inequality (Chung 2005):

$$2h(\overrightarrow{G}) \geq \lambda \geq \frac{h^2(\overrightarrow{G})}{2}. \tag{4.22}$$

For a strongly connected graph $\overrightarrow{G}$, the diameter $\mathrm{diam}(\overrightarrow{G})$ satisfies

$$\mathrm{diam}(\overrightarrow{G}) \leq 1 + \left\lfloor 2\frac{\min_i \ln(\pi_i^{-1})}{\ln \frac{2}{2-\lambda}} \right\rfloor, \tag{4.23}$$

where λ is the first nontrivial eigenvalue of the Laplace operator (4.15) and π is the Perron eigenvector of the random walk defined on $\overrightarrow{G}$ with the transition probability matrix (4.2).

4.4 Bi-Orthogonal Decomposition of Random Walks Defined on Strongly Connected Directed Graphs

In order to define the self-adjoint Laplace operator (4.15) on aperiodic strongly connected directed graphs, we need to know the stationary distributions of random walkers π. Even if π exists for a given directed graph $\overrightarrow{G}$, usually it can be evaluated only numerically in polynomial time (Lovasz et al. 1995). Stationary distributions on aperiodic general directed graphs are not so easy to describe since they are typically non-local in the sense that each coordinate π_i would depend upon the entire subgraph (the number of spanning arborescences of $\overrightarrow{G}$ rooted at i (Lovasz et al. 1995), but not on the local connectivity property of a node itself like it was in undirected graphs.

Furthermore, if the greatest common divisor of its cycle lengths in $\overrightarrow{G}$ exceeds 1, then the transition probability matrices (4.2) and (4.5) can have several eigenvectors belonging to the largest eigenvalue 1, so that the definition (4.15) of Laplace operator seems to be questionable.

4.4.1 Dynamically Conjugated Operators of Random Walks

Let us choose the natural counting measure μ_0 (2.17) defined on a strongly connected directed graph $\overrightarrow{G}$ specified by its adjacency matrix $\mathbf{A}_{\overrightarrow{G}} \neq \mathbf{A}_{\overrightarrow{G}}^{\top}$ such that $\deg_{\mathrm{in}}(i) \neq 0$ and $\deg_{\mathrm{out}}(i) \neq 0$ for $\forall i \in \overrightarrow{G}$, and consider two random walks operators.

The first operator represented by the matrix

$$\mathbf{P} = \mathbf{D}_{\text{out}}^{-1}\mathbf{A}_{\overrightarrow{G}}, \tag{4.24}$$

in which $\mathbf{D}_{\text{out}}$ is a diagonal matrix with entries $\deg_{\text{out}}(i)$, describes the time forward random walks of the nearest neighbor type defined on $\overrightarrow{G}$. Given a time forward vertex sequence $\mathfrak{w}$ rooted at $i \in \overrightarrow{G}$, the matrix element P_{ij} gives the probability that $j \in \overrightarrow{G}$ is the vertex next to i in $\mathfrak{w}$. Another operator is a dynamically conjugated operator to (4.24),

$$\begin{aligned} \mathbf{P}^{\star} &= \mathbf{D}_{\text{in}}^{-1}\mathbf{A}_{\overrightarrow{G}}^{\top} \\ &= \mathbf{D}_{\text{in}}^{-1}\mathbf{P}^{\top}\mathbf{D}_{\text{out}}, \end{aligned} \tag{4.25}$$

where $\mathbf{D}_{\text{in}} = \text{diag}\,(\deg_{\text{in}}(i))$. It describes random walks in time backward vertex sequences $\mathfrak{w}'$.

It is worth mentioning that being defined on undirected graphs $\mathbf{P}^{\star} \equiv \mathbf{P}$, since $\deg_{\text{in}}(i) = \deg_{\text{out}}(i)$ for $\forall i \in \overrightarrow{G}$ and $\mathbf{A}_G = \mathbf{A}_G^{\top}$. While on directed graphs, $\mathbf{P}^{\star}$ is related to $\mathbf{P}$ by the transformation

$$\mathbf{P} = \mathbf{D}_{\text{out}}^{-1}\left(\mathbf{P}^{\star}\right)^{\top}\mathbf{D}_{\text{in}}, \tag{4.26}$$

so that these operators are not adjoint, in general $\mathbf{P}^{\top} \neq \mathbf{P}^{\star}$.

4.4.2 Measures Associated with Random Walks

The measure associated with random walks defined on undirected graphs was specified by (2.23). Correspondingly, we can define two different measures

$$\mu_+ = \sum_j \deg_{\text{out}}(j)\delta(j), \quad \mu_- = \sum_j \deg_{\text{in}}(j)\delta(j) \tag{4.27}$$

associated with the *out-* and *in-*degrees of nodes of the directed graph. In accordance with (4.27), we also define two Hilbert spaces $\mathcal{H}_+$ and $\mathcal{H}_-$ associated consequently with the spaces of square summable functions, $\ell^2(\mu_+)$ and $\ell^2(\mu_-)$, by setting the norms as

$$\|x\|_{\mathcal{H}_{\pm}} = \sqrt{\langle x,x \rangle_{\mathcal{H}_{\pm}}},$$

where $\langle \cdot, \cdot \rangle_{\mathcal{H}_{\pm}}$ denotes the inner products with respect to measures (4.27). Then a function $f(j)$ defined on the set of graph vertices is $f_{\mathcal{H}_-}(j) \in \mathcal{H}_-$ if transformed by

$$f_{\mathcal{H}_-}(j) \to J_-f(j) \equiv \mu_{-j}^{-1/2}f(j) \tag{4.28}$$

and is $f_{\mathcal{H}_+}(j) \in \mathcal{H}_+$ while being transformed accordingly to

$$f_{\mathcal{H}_+}(j) \to J_+f(j) \equiv \mu_{+j}^{-1/2}f(j). \tag{4.29}$$

The obvious advantage of the measures (4.27) against the natural counting measure μ_0 is that the matrices of the transition operators P and $P^\star$ transformed under the change of measures as

$$P_\mu = J_+^{-1} P J_-, \quad P_\mu^\star = J_-^{-1} P^\star J_+, \tag{4.30}$$

become adjoint,

$$
\begin{aligned}
\left(P_\mu\right)_{ij} &= \frac{A_{\vec{G}\,ij}}{\sqrt{\deg_{\mathrm{out}}(i)}\,\sqrt{\deg_{\mathrm{in}}(j)}}, \\
\left(P_\mu^\star\right)_{ij} &\equiv \left(P_\mu^\top\right)_{ij} = \frac{A_{\vec{G}\,ij}^\top}{\sqrt{\deg_{\mathrm{in}}(i)}\,\sqrt{\deg_{\mathrm{out}}(j)}}.
\end{aligned}
\tag{4.31}
$$

It is also important to note that

$$P_\mu : \mathcal{H}_- \to \mathcal{H}_+ \quad \text{and} \quad \mathscr{P}_\mu^\star : \mathcal{H}_+ \to \mathcal{H}_-.$$

4.4.3 Biorthogonal Decomposition

There is a singular value dyadic expansion (biorthogonal decomposition (Aubry et al. 1991a, b) for the operator,

$$\mathbf{P}_\mu = \sum_{k=1}^N \Lambda_k \varphi_{ki} \psi_{ki} \equiv \sum_{k=1}^N \Lambda_k |\varphi_k\rangle\langle\psi_k|, \tag{4.32}$$

where $0 \le \Lambda_1 \le \ldots \Lambda_N$ and the functions $\varphi_k \in \mathcal{H}_+$ and $\psi_k \in \mathcal{H}_-$ are related by the Karhunen-Loève dispersion (Karhunen 1944, Loeve 1955),

$$\mathbf{P}_\mu \varphi_k = \Lambda_k \psi_k \tag{4.33}$$

satisfying the orthogonality condition:

$$\langle \varphi_k, \varphi_s \rangle_{\mathcal{H}_+} = \langle \psi_k, \psi_s \rangle_{\mathcal{H}_-} = \delta_{ks}. \tag{4.34}$$

Since the operators P_μ and $P_\mu^\top$ act between the alternative Hilbert spaces, it is insufficient to solve just one equation in order to determine their eigenvectors φ_k and ψ_k (Aubry 1993).

Instead, two equations need to be solved,

$$
\begin{cases}
\mathbf{P}_\mu\,\varphi &= \Lambda\psi, \\
\mathbf{P}_\mu^\top\,\psi &= \Lambda\varphi,
\end{cases}
\tag{4.35}
$$

or, equivalently,

$$\begin{pmatrix} 0 & \mathbf{P}_\mu \\ \mathbf{P}_\mu^\top & 0 \end{pmatrix} \begin{pmatrix} \varphi \\ \psi \end{pmatrix} = \Lambda \begin{pmatrix} \varphi \\ \psi \end{pmatrix}. \tag{4.36}$$

The latter equation allows for a graph-theoretical interpretation. The block anti-diagonal operator matrix in the left-hand side of (4.36) describes random walks defined on a bipartite graph. Bipartite graphs contain two disjoint sets of vertices such that no edge has both endpoints in the same set. However, in (4.36), both sets are formed by one and the same nodes of the original graph $\overrightarrow{G}$ on which two different random walk processes specified by the operators P_μ and $P_\mu^\top$ are defined.

It is obvious that any solution of the equation (4.36) is also a solution of the system

$$U\varphi = \Lambda^2 \varphi, \quad V\psi = \Lambda^2 \psi, \tag{4.37}$$

in which $U \equiv \mathbf{P}_\mu^\top \mathbf{P}_\mu$ and $V \equiv \mathbf{P}_\mu \mathbf{P}_\mu^\top$, although the converse is not necessarily true.

The self-adjoint nonnegative operators $U : \mathscr{H}_- \to \mathscr{H}_-$ and $V : H_+ \to \mathscr{H}_+$ share one and the same set of eigenvalues $\Lambda^2 \in [0, 1]$, and the orthonormal functions $\{\varphi_k\}$ and $\{\psi_k\}$ constitute the orthonormal basis for the Hilbert spaces $\mathscr{H}_+$ and $\mathscr{H}_-$, respectively. The Hilbert-Schmidt norm of both operators,

$$\operatorname{tr}\left(P_\mu^\top P_\mu\right) = \operatorname{tr}\left(P_\mu P_\mu^\top\right) = \sum_{k=1}^N \Lambda_k^2 \tag{4.38}$$

is the global characteristic of the directed graph.

Provided the random walks are defined on a strongly connected directed graph $\overrightarrow{G}$, let us consider the functions $\rho^{(t)}(k) \in [0, 1] \times \mathbb{Z}_+$ representing the probability for finding a random walker at the node k, at time t. A random walker located at the source node k can go to the destination node k' through either node and all paths are combined in superposition. Being transformed in accordance with (4.28) and (4.29), these functions take the following forms: $(\rho_k^{(t)})_{\mathscr{H}_-} = \mu_-^{-1/2}\rho^{(t)}(k)$ and $(\rho_k^{(t)})_{\mathscr{H}_+} = \mu_+^{-1/2}\rho^{(t)}(k)$. Then, the self-adjoint operators $U : \mathscr{H}_- \to \mathscr{H}_-$ and $V : \mathscr{H}_+ \to \mathscr{H}_+$ with the matrix elements

$$U_{kk'} = \frac{1}{\sqrt{\deg_{\text{out}}(k)\deg_{\text{in}}(k')}} \sum_{i \in \overrightarrow{G}} \frac{A_{\overrightarrow{G}ik}^\top A_{\overrightarrow{G}ik'}}{\sqrt{\deg_{\text{out}}(i)\deg_{\text{in}}(i)}},$$

$$V_{k'k} = \frac{1}{\sqrt{\deg_{\text{out}}(k')\deg_{\text{in}}(k)}} \sum_{i \in \overrightarrow{G}} \frac{A_{\overrightarrow{G}k'i} A_{\overrightarrow{G}ki}^\top}{\sqrt{\deg_{\text{out}}(i)\deg_{\text{in}}(i)}} \tag{4.39}$$

define the dynamical system

$$\begin{cases} \sum_{k \in \overrightarrow{G}} \left(\rho_k^{(t)}\right)_{\mathscr{H}_-} U_{kk'} = \left(\rho_{k'}^{(t+2)}\right)_{\mathscr{H}_-}, \\ \sum_{k' \in \overrightarrow{G}} \left(\rho_{k'}^{(t)}\right)_{\mathscr{H}_+} V_{k'k} = \left(\rho_k^{(t+2)}\right)_{\mathscr{H}_+}. \end{cases} \tag{4.40}$$

Following (Aubry 1993), we define a symmetry $(\mathfrak{S}, \widetilde{\mathfrak{S}})$ by the action of two operators $\mathfrak{S} : \mathscr{H}_- \to \mathscr{H}_-$ and $\widetilde{\mathfrak{S}} : \mathscr{H}_+ \to \mathscr{H}_+$ such that

$$P_\mu \mathfrak{S} = \widetilde{\mathfrak{S}} P_\mu, \quad P_\mu \mathfrak{S}^\top = \widetilde{\mathfrak{S}}^\top P_\mu, \tag{4.41}$$

The commutation of P_μ with the symmetry $(\mathfrak{S}, \widetilde{\mathfrak{S}})$ implies the commutation of U with the partial symmetry $\mathfrak{S}$,

$$U\mathfrak{S} = \mathfrak{S}U, \tag{4.42}$$

and the commutation of V with the symmetry $\widetilde{\mathfrak{S}}$,

$$V\widetilde{\mathfrak{S}} = \widetilde{\mathfrak{S}}V. \tag{4.43}$$

If $\mathfrak{S}$ and $\widetilde{\mathfrak{S}}$ constitute the unitary representations of the finite symmetry group, its irreducible representations can be used in order to classify structures of the directed graph $\overrightarrow{G}$.

4.5 Spectral Analysis of Self-Adjoint Operators Defined on Directed Graphs

The spectral properties of self-adjoint operators U and V that determine two Markov processes on strongly connected directed graphs and share the same set of nonnegative eigenvalues, $\Lambda^2 \in [0, 1]$, can be analyzed by the method of characteristic functions as we did in Sect. 3.4.

In particular, the self-adjoint operators U and V can be used in order to investigate the structure of both directed components of the graph by means of the PCA method (see Sect. 3.2) precisely as it was done with the undirected graphs in spectral graph theory.

The self-adjoint operators U and V describe correlations between flows of random walkers entering and leaving nodes in a directed graph. In the framework of the spectral approach, these correlations are labelled by the eigenvalues Λ_k^2, and those belonging to the largest eigenvalues are essential for coherence of traffic through the different components of the transport network.

The approach which we propose below for the analysis of coherent structures that arise in strongly connected directed graphs is similar to one used for the spatiotemporal analysis of complex signals in Aubry et al. (1991a) and Aubry (1993), and refers to the Karhuen-Loéve decomposition in classical signal analysis (Karhunen 1944).

In the framework of biorthogonal decomposition discussed in Aubry et al. (1991a), the complex spatiotemporal signal has been decomposed into orthogonal temporal modes called chronos, and orthogonal spatial modes called topos. Then the spectral analysis of the phase-space of the dynamics and the spatiotemporal intermittency in particular have naturally led to the notions of "energies" and "entropies" (temporal, spatial, and global) of signals.

Each spatial mode has been associated with an instantaneous coherent structure which has a temporal evolution directly given by its corresponding temporal mode. In view of that the thermodynamic-like quantities were used in order to describe the complicated spatiotemporal behavior of complex systems.

In the present subsection, we demonstrate that a somewhat similar approach can also be applied to the spectral analysis of directed graphs. Namely, that the morphological structure of directed graphs can be related to the quantities extracted from biorthogonal decomposition of random walks defined on them. Furthermore, in such a context, the temporal modes and spatial modes introduced in Aubry et al. (1991a) are the eigenfunctions of correlation operators U and V.

Although our approach can be viewed as a version of the signal analysis, it is fundamentally different from that in principle. Transition probability operators satisfy the probability conservation property by definition that, in general, is not the case for the spatiotemporal signals generated by complex systems.

All coherent segments of a directed graph participate independently in the Hilbert-Schmidt norm (4.38) of the self-adjoint operators U and V,

$$\mathfrak{E}(\overrightarrow{G}) = \sum_{k=1}^{N} \Lambda_k^2. \tag{4.44}$$

Borrowing the terminology from a theory of signals and Aubry et al. (1991a), we can call (4.44) energy, the only additive characteristic of the directed graph $\overrightarrow{G}$. While introducing the projection operators (in Dirac's notation) by

$$\mathbb{P}_k^{(+)} = |\psi_k\rangle\langle\psi_k|, \quad \mathbb{P}_k^{(-)} = |\varphi_k\rangle\langle\varphi_k|, \tag{4.45}$$

we can decompose the energy (4.44) into two components related to the Hilbert spaces $\mathcal{H}_+$ and $\mathcal{H}_-$:

$$\begin{aligned}
\mathfrak{E}(\overrightarrow{G}) &= \mathfrak{E}^{(+)}(\overrightarrow{G}) + \mathfrak{E}^{(-)}(\overrightarrow{G}) \\
&= \sum_{k=1}^{N} \Lambda_k^2 \mathbb{P}_k^{(+)} + \sum_{k=1}^{N} \Lambda_k^2 \mathbb{P}_k^{(-)}.
\end{aligned} \tag{4.46}$$

Using (4.44) as the normalizing factor, we can introduce the relative energy for each coherent structure of the directed graph by

$$\mathfrak{e}_k = \frac{\Lambda_k^2}{\mathfrak{E}(\overrightarrow{G})} \tag{4.47}$$

and, following Aubry et al. (1991a), define the global entropy of coherent structures in the graph $\overrightarrow{G}$ as

$$\mathfrak{H}(\overrightarrow{G}) = -\frac{1}{\ln N} \sum_{k=1}^{N} \mathfrak{e}_k \ln \mathfrak{e}_k, \tag{4.48}$$

which is independent on the graph size N due to the presence of normalizing factor $1/\ln N$ and, therefore, can be used in order to compare different directed graphs. The global entropy of the graph $\overrightarrow{G}$ is zero if all its nodes belong to one and the same coherent structure (i.e., only one eigenvalue $\Lambda_k^2 \neq 0$). In the opposite case, $\mathfrak{H}(\overrightarrow{G}) \to 1$ if most of eigenvalues Λ_k^2 are equal (multiple).

The relative energy (4.47) can also be decomposed into the $\mathscr{H}_+$- and $\mathscr{H}_-$-components:

$$e_k^{(\pm)} = \frac{\Lambda_k^2 \mathbb{P}_k^{(\pm)}}{\mathfrak{E}^{(\pm)}(\overrightarrow{G})}, \tag{4.49}$$

and then the partial entropies can be defined as

$$\mathfrak{H}^{(\pm)}(\overrightarrow{G}) = -\frac{1}{\ln N} \sum_{k=1}^{N} e_k^{(\pm)} \ln e^{(\pm)}{}_k. \tag{4.50}$$

To conclude, we have seen that any strongly connected directed graph $\overrightarrow{G}$ can be considered as a bipartite graph with respect to the in- and out-connectivity of nodes. The biorthogonal decomposition of random walks is then used in order to define the self-adjoint operators on directed graphs describing correlations between flows of random walkers which arrive at and depart from graph nodes. These self-adjoint operators share the nonnegative real spectrum of eigenvalues, but different orthonormal sets of eigenvectors. The standard principal component analysis can also be applied to directed graphs. The global characteristics of the directed graph and its components can be obtained from the spectral properties of the self-adjoint operators.

4.6 Self-Adjoint Operators for Interacting Networks

The biorthogonal decomposition can also be implemented in order to determine coherent segments of two or more interacting networks defined on one and the same set of nodes V, $|V| = N$.

Given two different strongly connected weighted directed graphs $\overrightarrow{G}_1$ and $\overrightarrow{G}_2$ specified on the same set of N vertices by the non-symmetric adjacency matrices $\mathbf{A}^{(1)}$, $\mathbf{A}^{(2)}$, which entries are the edge weights, $w_{ij}^{(1,2)} \geq 0$, then the four transition operators of random walks can be defined on both networks as

$$\mathbf{P}^{(\alpha)} = \left(\mathbf{D}_{\text{out}}^{(\alpha)}\right)^{-1} \mathbf{A}^{(\alpha)}, \quad \left(\mathbf{P}^{(\alpha)}\right)^{\star} = \left(\mathbf{D}_{\text{in}}^{(\alpha)}\right)^{-1} \mathbf{A}^{(\alpha)}, \quad \alpha = 1, 2, \tag{4.51}$$

where $\mathbf{D}_{\text{out}/\text{in}}$ are the diagonal matrices with the following entries:

$$\deg_{\text{out}}^{(\alpha)}(j) = \sum_{i,j \to i} w_{ji}^{(\alpha)}, \quad \deg_{\text{in}}^{(\alpha)}(j) = \sum_{i,i \to j} w_{ij}^{(\alpha)}, \quad \alpha = 1, 2. \tag{4.52}$$

We can define for different measures,

$$\begin{aligned}
\mu_-^{(1)} &= \Sigma_j \deg_{\text{out}}^{(1)}(j)\delta(j), \quad \mu_+^{(1)} = \Sigma_j \deg_{\text{in}}^{(1)}(j)\delta(j), \\
\mu_-^{(2)} &= \Sigma_j \deg_{\text{out}}^{(2)}(j)\delta(j), \quad \mu_+^{(2)} = \Sigma_j \deg_{\text{in}}^{(2)}(j)\delta(j).
\end{aligned} \tag{4.53}$$

and four Hilbert spaces $\mathscr{H}_{\pm}^{(\alpha)}$ associated with the spaces of square summable functions, $\ell^2\left(\mu_{\pm}^{(\alpha)}\right)$, $\alpha = 1,2$.

Then the transition operators $P_{\mu}^{(\alpha)} : \mathscr{H}_{-}^{(\alpha)} \to \mathscr{H}_{-}^{(\alpha)}$ and $\left(P_{\mu}^{(\alpha)}\right)^{\star} : \mathscr{H}_{+}^{(\alpha)} \to \mathscr{H}_{+}^{(\alpha)}$ adjoint with respect to the measures $\mu_{\pm}^{(\alpha)}$ are defined by the following matrices:

$$
\begin{aligned}
\left(P_{\mu}^{(\alpha)}\right)_{ij} &= \frac{A_{\overrightarrow{G}}^{(\alpha)}{}_{ij}}{\sqrt{k_{\text{out}}^{(\alpha)}(i)}\sqrt{k_{\text{in}}^{(\alpha)}(j)}}, \\
\left(P_{\mu}^{(\alpha)}\right)_{ij}^{\star} &= \frac{A_{\overrightarrow{G}}^{(\alpha)\top}{}_{ij}}{\sqrt{k_{\text{in}}^{(\alpha)}(i)}\sqrt{k_{\text{out}}^{(\alpha)}(j)}}.
\end{aligned}
\tag{4.54}
$$

The spectral analysis of the above operators requires that four equations be solved:

$$
\begin{cases}
\mathbf{P}_{\mu}^{(\alpha)}\varphi^{(\alpha)} = \Lambda^{(\alpha)}\psi^{(\alpha)}, \\
\mathbf{P}_{\mu}^{(\alpha)\top}\psi^{(\alpha)} = \Lambda^{(\alpha)}\varphi^{(\alpha)},
\end{cases}
\tag{4.55}
$$

where $\alpha = 1,2$ as usual.

Any solution $\{\varphi^{(\alpha)}, \psi^{(\alpha)}\}$ of the system (4.55), up to the possible partial isometries,

$$
\mathbf{G}^{(\alpha)}\varphi^{(\alpha)} = \psi^{(\alpha)},
\tag{4.56}
$$

also satisfies the system

$$
\begin{aligned}
\mathbf{P}_{\mu}^{(2)\top}\mathbf{P}_{\mu}^{(1)}\mathbf{P}_{\mu}^{(1)\top}\mathbf{P}_{\mu}^{(2)}\psi^{(2)} &= \left(\Lambda^{(1)}\Lambda^{(2)}\right)^2\psi^{(2)}, \\
\mathbf{P}_{\mu}^{(1)\top}\mathbf{P}_{\mu}^{(2)}\mathbf{P}_{\mu}^{(2)\top}\mathbf{P}_{\mu}^{(1)}\psi^{(1)} &= \left(\Lambda^{(1)}\Lambda^{(2)}\right)^2\psi^{(1)}, \\
\mathbf{P}_{\mu}^{(1)}\mathbf{P}_{\mu}^{(1)\top}\mathbf{P}_{\mu}^{(2)}\mathbf{P}_{\mu}^{(2)\top}\varphi^{(1)} &= \left(\Lambda^{(1)}\Lambda^{(2)}\right)^2\varphi^{(1)}, \\
\mathbf{P}_{\mu}^{(2)}\mathbf{P}_{\mu}^{(2)\top}\mathbf{P}_{\mu}^{(1)}\mathbf{P}_{\mu}^{(1)\top}\varphi^{(2)} &= \left(\Lambda^{(1)}\Lambda^{(2)}\right)^2\varphi^{(2)},
\end{aligned}
\tag{4.57}
$$

in which $\left(\Lambda^{(1)}\Lambda^{(2)}\right)^2 \in [0,1]$. Operators in the l.h.s of the system (4.57) describe correlations between flows of random walkers which go through vertices following the links of either network. Their spectrum can also be investigated by the methods discussed in the previous subsection.

It is convenient to represent the self-adjoint operators from the l.h.s. of (4.57) by the closed directed paths shown in the diagram in Fig. 4.1. Being in the self-adjoint products of transition operators, $P_{\mu}^{(\alpha)}$ corresponds to the flows of random walkers which depart from either network, and $P_{\mu}^{(\alpha)\top}$ is for those which arrive at the network α. From Fig. 4.1 , it is clear that the self-adjoint operators in (4.57) represent all possible closed trajectories visiting both networks $\mathbf{N}_1$ and $\mathbf{N}_2$.

In general, given a complex system consisting of $n > 1$ interacting networks operating on the same set of nodes, we can define 2^n self-adjoint operators related to the different modes of random walks. Then the set of network nodes can be sepa-

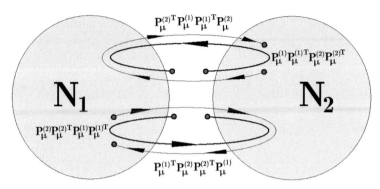

Fig. 4.1 Self-adjoint operators for two interacting networks sharing the same set of nodes

rated into a number of essentially correlated segments with respect to each of the self-adjoint operators.

4.7 Summary

Spectral methods have not been applied to directed networks because it is not always possible to uniquely define a linear self-adjoint operator on them.

We have implemented the method of biorthogonal decomposition on strongly connected directed graphs in order to define two self-adjoint operators - one for the in-component and another for the out-component of the graph. They can be interpreted as the operators of time-forward and time-backward random walks. The spectral properties of these self-adjoint operators can be used to analyze the directed networks.

A similar approach can be applied for interacting networks. For $n > 1$ interacting networks, we can define 2^n self-adjoint operators subjected to the standard spectral analysis developed in the preceding chapters.

Chapter 5
Urban Area Networks and Beyond

5.1 Miracle of Complex Networks

The ultimate goals of network theory are to help people in planning the efficient land use, feasible logistic, expeditious energy grids, and streamlined communications. Most of the life-sustaining networks are essentially small – usually, they amount to just several hundred nodes connected by no means at random.

Systems consisting of many individual units sharing information, people, and goods need to be modified frequently to meet the natural challenges. However, since the possible response of the entire network to the modifications is poorly understood, it has always been difficult to make the best decisions that will sustain the system and cope with the new demand. The counterintuitive phenomenon known as the Braess paradox is an example (Braess 1968); it occurs when adding more resources to a transport network (say, a new road or a bridge) deteriorates the quality of traffic by creating worse delays for drivers, rather than alleviating them. The Braess paradox has been observed in the street vehicular traffic of New York City and Stuttgart (Kolata 1990).

It appears that the networking is structurally "contagious." In order to master a network effectively, its authority should also constitute a network, probably as complex as one to be supervised. The governance and maintenance units supporting urban network renewals, negotiating performance targets, taking decisions, financing and eventually implementing them also form networks. In fact, a complex network of city itineraries that we can experience in everyday life arises as a net result of multiple complex interactions between a number of transport, social, and economic networks.

The concept of equilibrium, the condition for a system in which competing influences are balanced, is a key theoretical element in any branch of science. It is intuitively clear that a complex network while at equilibrium constitutes a dynamic form of synergy between the *topological* structure shaped by a connected graph G with some positive measures (or masses) appointed to the vertices and some positive weights assigned to the edges; the *dynamics* described by the set of operators defined on the graph G and the properties of embedding *physical space*. A change

Ph. Blanchard, D. Volchenkov, *Mathematical Analysis of Urban Spatial Networks,*
Understanding Complex Systems, DOI 10.1007/978-3-540-87829-2_5,
© Springer-Verlag Berlin Heidelberg 2009

made to any component of this "saint trinity" would have a great effect on the entire complex network, definitely moving it out off equilibrium and even causing the equilibrium state itself to dissolve.

Occasional improvements made from time to time to certain modes of transportation in the city might unwittingly create the hidden areas of geographical isolation in the urban landscape. Neighborhoods which are effectively cut off from other parts of the city by poor transport links and haphazard urban redevelopments often suffer from social ills fostering crime, deprivation and ghettoization (Chown 2007). It has been proven within the space syntax approach that isolation worsens an area's economic prospects by reducing opportunities for commerce, and engenders a feeling of isolation in inhabitants, both of which can fuel poverty and crime.

Unfortunately, the business and development administrations have often failed to take such isolation into account when shaping the city landscape because isolation can sometimes be difficult to quantify in the complex fabric of a major city.

5.2 Urban Sprawl – a European Challenge

The Physical pattern of low-density expansion of large urban areas mainly into surrounding agricultural areas is described by the European Environment Agency (EEA) as sprawl (EEA 2006).

Urban sprawl has accompanied the growth of urban areas across Europe over the past 50 years. The most visible impact of urban sprawl arises in countries with high population density and economic activity as well as in regions with rapid economic growth. Sprawl is particularly evident around smaller towns, in the countryside, along transportation corridors, and along many parts of the coast usually connected to river valleys (EEA 2006).

It is well-known that urban development has impacts far beyond the land consumed directly by construction and infrastructure. Many environmental problems generated by the expansion of cities create dramatic economic and social implications for their cores with negative effects on the urban economy. All evidence shows that degraded urban areas are less likely to attract investments, new enterprises and services, but become attractive for socially underprivileged groups due to reduced house prices in the urban core (EU Environment 2006).

Urban sprawl involves the substantial consumption of numerous natural resources, such as scarce land resources which are mostly nonrenewable, that dramatically transforms the properties of soil, reducing its capacity to perform its essential functions.

A further consequence of the increasing consumption of land and reduction in population densities is the growing consumption of natural resources (as water) and energy per capita. Transport-related energy consumption also increases, dominated by relatively energy inefficient car use, while public transportation systems become increasingly expensive and typically inadequate. Sprawl-related growth of urban transport and greenhouse gas emissions have major implications for global warming and climate change as well as for the quality of life and human health in cities (Norman et al. 2006).

From a social perspective urban sprawl induces segregation of residential developments according to income and age. The young and old, who lack mobility and resources have poor chances for social interactions.

"The issue of mobility, and accessibility, therefore remains a critical challenge for urban planning and management, as well as a key factor in European territorial cohesion" (EEA 2006).

Despite the complexity of urban systems, sprawl is seldom tackled as an integrated issue. Local urban and regional management often respond to urban sprawl by creating parallel or shortcut routes connecting the different parts of sprawl. It is believed that shortcuts could improve the accessibility to certain suburban areas and alleviate the problem of traffic in the entire city.

We would like to stress that the planning of shortcuts is a highly nontrivial task requiring a thorough analysis and impact assessment; nevertheless it probably does not provide the optimal solutions for either the traffic problems, nor the city's connectedness and navigation challenges.

Let us consider the relatively small urban pattern of Neubeckum as an example. In Fig. 5.1, we have represented four different shortcuts (marked on the city plan by the letters A, B, C, and D, respectively) that can be created between sprawling developments and main city itineraries and investigated their implications for the accessibility of two reference points in the city (the mosque and the church). For all proposed shortcuts (A,B,C, and D) and various combinations of them (AB, ABD, BCD, and ABCD), we have estimated relative isolation (quantified by the

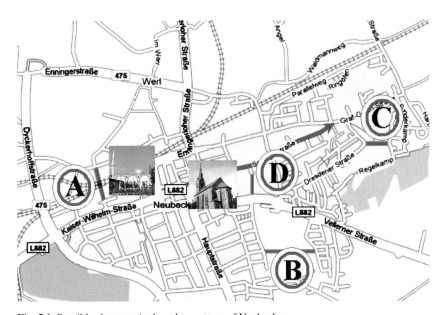

Fig. 5.1 Possible shortcuts in the urban pattern of Neubeckum

first-passage times) of the reference points, with respect to the best accessible place of motion in Neubeckum – Spiekersstrasse, the diameter $\mathscr{D}$ of the spatial graph of Neubeckum, and the structural distance dist(Church, Mosque) between the locations of the church and the mosque estimated by the commute time, the expected number of steps a random walker needs to reach the mosque from the church and then to return.

Results on the structural analysis of shortcuts in the urban pattern of Neubeckum is represented in Table 5.1. They demonstrate convincingly that

- while improving local accessibility in the nearest neighborhood, a shortcut may have negative implications on the accessibility level of faraway neighborhoods as well as on the entire connectedness of the city;
- planning of shortcuts does not show any "cumulative effect:" two "good" shortcuts which separately would improve accessibility to the individual neighborhoods locally, may worsen isolation if created simultaneously;
- creation of scattered local shortcuts does not help much in giving a compact form to the city.

We conclude that urban sprawl must be considered in connection with the main urban area. In search of a compact structural pattern lost due to the imperceptible flow of urban developments across municipal boundaries, one might consider a large-scale replanning of the city structure including its core. One of the best practice experiences highlighting the success of the city planners can be found in the Munich area which remains exceptionally compact, compared to many other European cities. The post-war decision of the planners to rebuild the historical center and replace traditional town planning with integrated urban development planning provides a guideline to the management of urban sprawl throughout Europe (EEA 2006).

Table 5.1 Structural analysis of shortcuts in the urban pattern of Neubeckum

	Church	Mosque	D	dist(Church, Mosque)
	FPT(steps)/I(dB)	FPT(steps)/I(dB)		(steps)
Present	40/3.01	345/12.37	7	405
A	41/3.11	202/10.04	7	271
B	39/2.69	350/12.22	7	409
C	35/2.22	356/12.29	7	412
D	34/3.01	348/12.29	6	398
AB	40/2.80	205/10.88	7	274
ABD	33/2.88	208/10.87	6	265
BCD	33/2.18	367/12.49	6	404
ABCD	36/2.98	212/10.97	6	317

5.3 Ranking Web Pages, Web Sites, and Documents

The huge business success of Google has drawn meticulous attention to the models of the query independent web graph structure analysis (Brin et al. 1999, Kleinberg 1999). The success of web search algorithms emerges when the individual expectations of a large number of internet users surfing the web in search of required data are mostly enjoyed by prompt and adequate responses from a search engine.

Web sites are semantically important for understanding the whole picture of the World Wide Web, and the ranking of them is a key technical component in many commercial search engines. Interestingly, web sites contain a higher density (about 75%) of hyperlinks (Henzinger et al. 2003), while the fraction of edges in between sites is typically much lower (Girvan et al. 2002). The inter-connectivity between web sites is described by the Host Graph in which web sites are nodes and interlinks are edges connecting them.

Random walks have been used in order to investigate the properties of the Host Graph and to calculate the web site ranking by Bharat et al. (2001) and Dill et al. (2001). Bao et al. (2006) suggested evaluating the importance of a web site with the mean frequency of its visiting by a general Markov chain defined on the internet graph.

Algorithms described in the previous chapters can be directly implemented to understand the World Wide Web structure and web site ranking. In particular, it is important to note that an algorithm based on the estimation of the first-passage time to a web site precisely reflects the strategy of crawler programs used by commercial search companies while collecting their databases.

Indeed, the web surfing strategy followed by a particular internet user is by no means random, but is strongly skewed with the individual preferences and aptitudes and essentially biased by an internet service provider. The conspicuous structural discrepancy between the strongly biased searching algorithms used by humans and the random walks like technics typically implemented by crawlers classifying documents would lead to an ineffective and even erroneous search.

The comparative analysis of the first-passage properties of documents and web sites in biased and unbiased frames can help to estimate and avoid errors in searching algorithms.

5.4 Image Processing

Many recent techniques for digital image enhancement and multiscale image representations are based on nonlinear partial differential equations which describe anisotropic diffusion processes (Weickert 1998). However, when applied to real image processing the semidiscrete and fully discrete settings are more important. The concept of image processing by use of self-organization mechanisms in a discrete nonlinear system has been proposed by Ebihara et al. (2003) where edge detection, image segmentation, noise reduction and contrast enhancement was achieved using a discrete reaction-diffusion model (FitzHugh-Nagumo model) under the condition of Turing instability. Compared with the conventional method, the proposed technique indicates a higher performance in processing for noisy images.

Different types of diffusion distances have been proposed independently by Bloch (1999), Ling et al. (2006), and by many other authors to measure dissimilarity between histogram-based descriptors. The first-passage time algorithms may be used in order to estimate similarity between histogram-based descriptors and to reduce image sizes. Indeed, while processing an image, we can embed it into metric Euclidean space, in which processing errors may be estimated precisely. We have no right to reduce the number of pixels in the image; however, we can dramatically cut down the number of diffusion eigenmodes defined on them by constructing a low-dimensional representation of the image matrix (see Chap. 3) sustaining the optimal image representation quality.

In Fig. 5.2, we have presented the result of implementating the described algorithm, consequently reducing the number of used eigenmodes in the image of Sir Isaac Newton.

The first-passage time distance performs excellently in both accuracy and efficiency, compared with other state-of-the-art distance measures.

Fig. 5.2 Sir Isaac Newton's image is consequently represented with (**1**) 300 eigenmodes; (**2**) 15 eigenmodes; (**3**) 5 eigenmodes; (**4**) 2 eigenmodes (*black* and *white*).

5.5 Summary

Occasional improvements made from time to time to certain routes or modes of transportation in the city might unwittingly provoke isolation in the urban landscape. While investigating urban sprawl, we have concluded that it must be considered in connection with the main urban area and that a large-scale re-planning may be necessary in order to control urban sprawl.

We have also shown that algorithms based on the estimation of the first-passage time may be used for ranking web pages, web sites, and documents, as well as in image processing.

Bibligraphy

M. Abramowitz, I. A. Stegun, eds. *Handbook of Mathematical Functions with Formulas, Graphs, and Mathematical Tables*, Dover (1972).

R.K. Ahuja, T.L. Magnanti, J.B. Orlin, T.L. Magnanti, *Network Flows: Theory, Algorithms, and Applications*, Prentice Hall (1993).

R. Albert, A.L. Barabási, "Statistical mechanics of complex networks". *Rev. Mod. Phys.* **74**, 47 (2002).

D.J. Aldous, J.A. Fill, *Reversible Markov Chains and Random Walks on Graphs*. In preparation (2008), available at www.stat.berkeley.edu/aldous/book.html.

G. Alexanderson, "Euler and Königsberg's bridges: a historical view". *Bullet. Am. Math- Soc* **43**, 567–573 (2006).

N. Alon, M. Krivelevich, V. H. Vu, " On the concentration of eigenvalues of random symmetric matrices". *Israel J. Math.* **131**, 259–267, (2002).

R.A. Andersen, G.K. Essick, R.M. Siegel, "Encoding of spatial location by posterior parietal neurons." *Science* **230**, 456–458 (1985).

H.R. Anderson, *Fixed Broadband Wireless System Design*, Wiley, New York (2003).

C. Andersson, A. Hellervik. K. Lindgren, "A spatial network explanation for a hierarchy of urban power laws". *Physica A* **345**, 227 (2005).

L. Arnold, "On Wigner's Semicircle Law for the Eigenvalues of Random Matrices." *Z. Wahrscheinlichkeitstheorie und Verw. Gebiete* **19**, 191–198, (1971).

N. Aubry, R. Guyonnet, R. Lima, "Spatiotemporal analysis of complex signals: theory and applications". *J. Stat. Phys.* **64**, 683–739 (1991).

N. Aubry, "On the hidden beauty of the proper orthogonal decomposition". *Theor. Comp. Fluid Dyn.* **2**, 339–352 (1991).

N. Aubry, L. Lima, *Spatio-temporal symmetries*, Preprint CPT-93/P.2923, Centre de Physique Theorique, Luminy, Marseille, France (1993).

F. Auerbach, "Das Gesetz der Bevölkerungskonzentration", *Petermanns Geographische Mitteilungen* **59**, 74-76 (1913).

F.R. Bach, M.I. Jordan, "Learning spectral clustering", *Technical report, UC Berkeley*, available at www.cs.berkeley.edu/fbach (2003); Tutorial given at *ICML 2004 International Conference on Machine Learning*, Banff, Alberta, Canada (2004).

B.T. Backus, I. Oruç, "Illusory motion from change over time in the response to contrast and luminance". *Jour. Vision,* **5**(11):10, 1055–1069, doi:10.1167/5.11.10 (2005).

J.P. Bagrow, E.M. Bollt, J.D. Skufca, D. ben-Avraham, "Portraits of complex networks", *Eur. Phys. Lett.* **81**, 68004 (6pp) (2008); doi:10.1209/0295-5075/81/68004.

Y. Bao, G. Feng, T.-Y. Liu, Z.-M. Ma, Y. Wang, "Ranking websites: A probabilistic view", *Internet Math.* **3** (3), 295–320 (2006).

A.-L. Barabási, R. Albert, "Emergence of Scaling in Random Networks". *Science* **286**, 509–512 (1999).

A.-L. Barabàsi, *Linked: How Everything is Connected to Everything Else,* Penguin, (2004).

A.-L. Barabási, "The origin of bursts and heavy tails in human dynamics." *Nature* (London) **435**, 207–211 (2005).

M. Batty, P. Longley, *Fractal Cities,* Academic Press. London (1994).

A. Beck, M. Bleicher, D. Crowe, *Excursion into Mathematics,* Worth Publishers (1969).

F. Benford, "The law of anomalous numbers", *Proc. Am. Philos. Soc.* **78**, 551–572 (1938).

L.M.A. Bettencourt, J. Lobo, D. Helbing, C. Kühnert, G.B. West, "Growth, innovation, scaling, and the pace of life in cities". *PNAS* **104** (17), 7301–7306 (2007); doi:10.1073/pnas.0610172104.

J.M.H. Beusmans, "Computing the direction of heading from affine image flow". *Biol. Cybernetics* **70**, 123–136 (1993).

K. Bharat, B.-W. Chang, M. Henzinger, M. Ruhl, "Who links to whom: Mining linkage between web sites". *Proc. IEEE Int. Conf. Data Mining* ICDM'01, San Jose, USA (2001).

G. Bianconi, N. Gulbahce, A.E. Motter, *Local structure of directed networks,* Phys. *Rev. Lett.* **100**, 118701 (2008).

N. Biggs, *Algebraic Graph Theory,* Cambridge University Press (1974).

N. Biggs, *Permutation groups and combinatorial structures,* Cambridge University Press (1979).

N. Biggs, E. Lloyd, R. Wilson, *Graph Theory, 1736–1936.* Oxford University Press (1986).

S. Bilke, C. Peterson, "Topological properties of citation and metabolic networks". *Phys. Rev. E* **64**, 036106 (2001).

T. Biyikoğlu, W. Hordijk, J. Leydold, T. Pisanski, P.F. Stadler, "Graph laplacians, nodal domains, and hyperplane arrangements". *Lin. Alg. Appl.* **390**, 155–174 (2004).

A. Björner, L. Lovász, P. Shor, "Chip-firing games on graphs." *Eur. J. Combin.* **12**, 283–291 (1991).

A. Björner, L. Lovász, "Chip-firing games on directed graphs." *J. Algebraic Comb.* **1**, 305–328 (1992).

Ph. Blanchard, T. Krüger, "The *Cameo Principle* and the origin of scale-free graphs in social networks", *J. Stat. Phys.* **114** (5–6), 1399–1416 (2004).

Ph. Blanchard, M.-O. Hongler, "Modeling human activity in the spirit of Barabasi's queueing systems". *Phys. Rev. E* **75**, 026102 (2007).

Ph. Blanchard, D. Volchenkov, "Intelligibility and first passage times in complex urban networks". *Proc. R. Soc. A* **464**, 2153–2167 doi:10.1098/rspa.2007.0329 (2008).

I. Bloch, "On fuzzy distances and their use in image processing under imprecision" *Pattern recognition* **32** (11), 1873–1895, ISSN 0031-3203 CODEN PTNRA8 (1999).

S. Boccaletti, V. Latora, Y. Moreno, M. Chavez, D.-U. Hwang, "Complex networks: Structure and dynamics". *Phys. Rep.* **424**, 175 (2006).

G. Bolch, S. Greiner, H. de Meer, K. S. Trivedi, *Queueing Networks and Markov Chains: Modeling and Performance Evaluation with Computer Science Applications*, John Wiley, second edition, New York, NY (2006).

B. Bollobas, *Graph Theory*, Springer (1979).

B. Bollobás, O. Riordan, "Mathematical results on scale-free random graphs", in *Handbook of Graphs and Networks*, Wiley-VCH (2002).

P.H.L. Bovy, E. Stern, *Route Choice: Wayfinding in Transport Networks*, Kluwer Academic Publishers, Dordrecht (1990).

D. Braess, "Über ein Paradoxon aus der Verkehrsplannung". *Unternehmensforschung* **12**, 258–268 (1968).

S. Brakman, H. Garretsen, C. van Marrewijk, M. van den Berg, *J. Reg. Sci.* **39**, 183–213 (1999).

S. Brakman, H. Garretsen, C. van Marrewijk, *An Introduction to Geographical Economics*, Cambridge Univ. Press, Cambridge, New York (2001).

A. Bretagnolle, E. Daudé, D. Pumain, "Cybergeo", *Revue Européene de Géographie* **335**, 1 (2006).

A. Brettel, *The effects of "order" and "disorder" on human cognitive perception in navigating through urban environments*. Thesis (Masters.MSc), UCL (University College London) (2006).

L. Breuer, D. Baum, *An Introduction to Queueing Theory*, Springer (2005).

S. Brin, L. Page, R. Motwami, T. Winograd, "The pageRank citation ranking: Bringing order to the web". *Tech. Rep. 1999-0120*, Computer Science Dept., Stanford University, Stanford, CA (1999).

N.J. Broadbent, L.R. Squire, R.E. Clark, "Spatial memory, recognition memory, and the hippocampus". *PNAS* **101** (40), 14515–14520 (2004).

E. Brockmeyer, H.L. Halstrøm, A. Jensen, "The life and works of A.K. Erlang". In *Transactions of the Danish Academy of Technical Sciences* **2**. Available at http://oldwww.com.dtu.dk/teletraffic/Erlang.html (1948).

R. Brookhiser, "Urban Sundial - Manhattan's grid system streets", *Nati. Rev.*, NY, July 9 (2001).

M. Buchanan, *Nexus: Small worlds and the Groundbreaking Theory of Networks*. Norton, W. W. and Co., Inc. ISBN 0-393-32442-7 (2003).

D. Burago, Yu.D. Burago, S. Ivanov, *A Course in Metric Geometry*, AMS (2001); ISBN 0-8218-2129-6.

H. Busemann, P.J. Kelly, *Projective Geometry and Projective Metrics*, in *Pure and Applied Mathematics* **3** (eds.) P.A. Smith, S. Eilenberg, Academic Press Inc., Publishers (1953).

S. Butler, *Spectral graph theory*, 3 lectures given at the Center for Combinatorics, Nankai University, Tianjin, September (2006).

S. Butler, "Interlacing for weighted graphs using the normalized Laplacian". *Electronic J. Lin. Alg.* **16**, 90-98 (2007).

A. Cardillo, S. Scellato, V. Latora, S. Porta, "Structural properties of planar graphs of urban street patterns". *Phys. Rev. E* **73**, 066107 (2006).

R. Carvalho, A. Penn, "Scaling and universality in the micro-structure of urban space". *Physica A*, **332**, 539–547 (2004).

M. Castells, *The Internet Galaxy: Reflections on the Internet, Business, and Society*, Oxford (2001).

L.L. Cavalli-Sforza, *Genes, Peoples, and Languages*, North Point Press (2000).

P. Chan, M. Schlag, J. Zien, *IEEE Trans. CAD-Integrated Circuits and Systems* **13**, 1088–1096 (1994).

J. Cheeger, "A lower bound for the smallest eigenvalue of the Laplacian", *Problems in Analysis, Papers dedicated to Salomon Bochner*, Princeton University Press 195–199 (1969).

K. Cheng, "A purely geometric module in the rat's spatial representation". *Cognition* **23**, 149–178 (1986).

M. Chown, "The future poverty hiding in cities". *New Scient* **2628**, 3 November 2007.

W. Christaller, *Central Places in Southern Germany*, Prentice Hall, Englewood Cliffs, NJ (1966).

F. Chung, "Diameters and eigenvalues". *J. Am. Math. Soc.* **2**, 187–196 (1989).

F. Chung, R.L. Graham, R.M. Wilson, "Quasi-random graphs". *Combinatorica* **9**, 345–362 (1989).

F. Chung, *Lecture notes on spectral graph theory*, AMS Publications Providence (1997).

F. Chung, L. Lu, V. Vu, "Spectra of random graphs with given expected degrees." *Proc. Natl. Acad. Sci. USA.* **100**(11): 6313–6318. (2003); Published online doi: 10.1073/pnas.0937490100.

F. Chung, "Laplacians and the Cheeger inequality for directed graphs," *Annals of Combinatorics* **9**, 1–19 (2005).

Churchill, Famous Quotations/Stories of Winston Churchill at http://www. winstonchurchill.org.

J. Cohen, P. Cohen, S.G. West, L.S. Aiken, *Applied Multiple Regression /Correlation Analysis for the Behavioral Sciences.* (3rd ed.) Hillsdale, NJ: Lawrence Erlbaum Associates (2003).

D. Coppersmith, P. Tetali, P. Winkler. "Collisions among random walks on a graph". *SIAM J. Discrete Math.* **6**(3), 363-374 (1993).

T.H. Cormen, C.E. Leiserson, R.L. Rivest, C. Stein. *Introduction to Algorithms*, Second Edition. MIT Press and McGraw-Hill, (2001). ISBN 0-262-03293-7. Chapter 21: Data structures for Disjoint Sets, pp.498-524.

L. da F. Costa, G. Travieso, "Exploring complex networks through random walks", *Phys. Rev. E* **75**, 016102 (2007).

T.M. Cover, J.A. Thomas, *Elements of Information Theory.* New York: Wiley, (1991).

P. Crucitti, V. Latora, S. Porta, "Centrality in networks of urban streets". *Chaos* **16**, 015113 (2006).

C. R. Dalton, N.S. Dalton, "A spatial signature of sprawl: or the proportion and distribution of linear network circuits". In: *GeoComputation 2005*, 1–3 August 2005, Ann Arbor, Michigan, USA (2005).

C.R. Dalton, N.S. Dalton, "The theory of natural movement and its application to the simulation of mobile ad hoc networks (MANET)". In 5th *Annual Conference on Communication Networks and Services Research (CNSR2007)*, 14-17 May 2007, New Brunswick, Canada (2007).

G.B. Dantzig, R. Fulkerson, S. M. Johnson, "Solution of a large-scale. traveling salesman problem". *Operations Research*, **2**, 393–410 (1954).

E.B. Davis, G.M.L. Gladwell, J. Leydold, P.F. Stadler, "Discrete nodal domain theorems". *Lin. Alg. Appl.* **336**, 51–60 (2001).

R. Diestel, *Graph Theory*, Springer (2005).

M. Deyer, A. Frieze, R. Kannan, "A random polynomial time algorithm for estimating volumes of convex bodies", *Proc. 21st Annual ACM Symposium on the theory of Computing*, 68–74 (1986).

I. Dhillon, Y. Guan, B. Kulis, "A unified view of kernel k-means, spectral clustering and graph cuts". *Technical Report TR-04-25*, University of Texas at Austin (2004).

I. Dhillon, Y. Guan, B. Kulis, "Kernel k-means: spectral clustering and normalized cuts", in *Proceedings of the 10th ACM SIGKDD international conference on Knowledge discovery and data mining*, Seattle, WA, USA (2004).

P. Diaconis, *Group Representations in Probability and Statistics*, Inst. of Math. Statistics, Hayward, CA (1988).

S. Dill, R. Kumar, K. McCurley, S. Rajagopalan, D. Sivakumar, A. Tomkins, "Self-similarity in the Web." *Proc. Int. Conf. Very Large Data Bases*, 69–78 (2001).

C. Ding, X. He, "K-means clustering via principal component analysis". *Proc. of Intl. Conf. Machine Learning (ICML'2004)*, 225–232. July 2004.

L. Donetti, F. Neri and M. A. Muñoz, "Optimal network topologies: Expanders, Cages, Ramanujan graphs, Entangled networks and all that." *J. Stat. Mech.*, P08007 (2006), doi:10.1088/1742-5468/2006/08/P08007.

S.N. Dorogovtsev, A.V. Goltsev, J.F.F. Mendes, A.N. Samukhin, "Renormalization group for evolving networks." *Phys. Rev. E* **68**, 046109 (2003).

C. Dunn, "Crime area research". In: D. Georges, K. Harris, (eds.), *Crime: A Spatial Perspective*, NY, Columbia Press (1980).

E. Durkheim, *The Division of Labor in Society*, Free Press, (1964).

M. Ebihara, H. Mahara, T. Sakurai, A. Nomura, H. Miike, "Image Processing by a Discrete Reaction-diffusion System", from *Proc. Visualization, Imaging, and Image Processing* (2003).

EEA, European Environment Agency report *Urban sprawl in Europe. The ignored challenge*, ISBN 92-9167-887-2 (2006).

M.Egenhofer, D.M. Mark, *Naive Geography* In: A.U. Frank, W. Kuhn (eds.) *Spatial Information Theory: A Theoretical Basis for GIS*, Springer, (1995).

G.B. West, J.H. Brown, B.J. Enquist, "A general model for the origin of allometric scaling laws in biology." *Nature* **395**, 163–166 (1998).

P. Erdös, A. Renyi, "On random graphs" *Publ. Math. Debrecen* **6**, 290 (1959).

K.A. Eriksen, I. Simonsen, S. Maslov, K. Sneppen, "Modularity and extreme edges of the interet". *Phys. Rev. Lett.* **90** (14), 148701 (2003).

EU Environment 2006 "Improving the quality of life in urban areas - Investments in awareness raising and environmental technologies". The Discussion paper for the *Informal Meeting of EU Environment Ministers*, 19–20 May 2006 Eisenstat/Rust. Austria.

A. Fabrikant, E. Koutsoupias, C.H. Papadimitriou. *Heuristically optimized trade-offs: A new paradigm for power laws in the internet*. ICALP (2002).

C. Faloutsos, M. Faloutsos, P. Faloutsos, *On power-law relationships of the internet topology*, in *Proc. CIGCOMM* (1999).

K. Fan, "On a theorem of Weyl concerning eigenvalues of linear transformations." *Proc. Natl. Acad. Sci. USA* **35**, 652–655 (1949).

I.J. Farkas, I. Derényi, A.-L. Barabási, T. Vicsek, "Spectra of "Real-World" graphs: Beyond the semi-circle law." *Phys. Rev. E* **64**, 026704 (2001).

I. Farkas, I. Derényi, H. Jeong, Z. Neda, Z.N. Oltvai, E. Ravasz, A. Schubert, A.-L. Barabási, T. Vicsek, "Networks in life: Scaling properties and eigenvalue spectra". *Physica A* **314**, 25 (2002).

O. Faugeras, "Stratification of three-dimensional vision: projective, affine, and metric representations". *J. Opt. Soc. Am. A* **12**(3), 465–484 (1995).

M. Fiedler, Algebraic connectivity of graphs." *Czech. Math. J.* **23**, 298 (1973); **25**, 146 (1975).

L. Figueiredo, L. Amorim, *Continuity lines in the axial system*, in A. Van Nes (ed.): 5th *International Space Syntax Symposium, TU Delft, Faculty of Architecture, Section of Urban Renewal and Management*, Delft, pp. 161–174 (2005).

L. Figueiredo, L. Amorim, *Decoding the urban grid: or why cities are neither trees nor perfect grids*, 6th International Space Syntax Symposium, 12–15 Jun 2007, Istanbul, Turkey.

M. Fisher, "Urban ecology". *Permaculture Design course handout notes* available at *www.self − willed − land.org.uk* (2008).

M. Flanders, S. Helms Tillery, J.F. Soechting, "Early stages in a sensorimotor transformation", *Behavioral and Brain Sciences* **15**, 309–362 (1992).

R. Florida, *Cities and the Creative Class*, Routledge, New York (2004).

M. Frigo, C.E. Leiserson, H. Prokop, "Cache-oblivious algorithms", in *40th Annual Symposium on Foundations of Computer Science, FOCS '99*, New York, NY, USA. IEEE Computer Society (1999).

K. Fukunaga, *Introduction to Statistical Pattern Recognition*, ISBN 0122698517, Elsevier (1990).

S. Funahashi, C.J. Bruce, P.S. Goldman-Rakic, "Mnemonic coding of visual space in the monkey's dorsolateral prefrontal cortex." *J. Neuropysiol.* **61**, 331–349 (1989).

M. Fyhn, T. Hafting, A. Treves, M.-B. Moser, E.I. Moser, "Hippocampal remapping and grid realignment in entorhinal cortex." *Nature* **446**, 190–194 (2007).

X. Gabaix, "Zipf's Law for Cities: An Explanation," *Quart. J. Econ.*, **113** (3), 739–767 (1999).

X. Gabaix, Y.M. Ioannides, "The evolution of city size distributions," in: J.V. Henderson, J.F. Thisse (ed.), *Handbook of Regional and Urban Economics*, (ed. 1, vol. **4**), ch. 53, 2341–2378 Elsevier (2004).

GEO-4: environment for development. United Nations Environment Program, ISBN: 978-92-807-2836-1 (UNEP paperback) DEW/0962/NA, Progress Press Ltd, Malta (2007).

M. Girvan, M.E.J. Newman, "Community structure in social and biological networks". *PNAS* 7821–7826 (2002).

C. Godsil, G. Royle, *Algebraic graph theory.* New York Springer-Verlag (2004). ISBN 0-387-95241-1.

R.G. Golledge, *Wayfinding Behavior: Cognitive Mapping and Other Spatial Processes*, John Hopkins University Press, ISBN: 0-8018-5993-X (1999).

G. Golub, C. van Loan,*Matrix computations*, 3rd edition, The Johns Hopkins University Press, London (1996).

J. Gomez-Gardenes, V. Latora, "Entropy rate of diffusion processes on complex networks". E-print arXiv:0712.0278v1 [cond-mat.stat-mech] (2007); also in V. Latora, *Proceedings of the Workshop Net-Works 2008*, Pamplona, 9–11 June, (2008).

A. Graham, *Nonnegative Matrices and Applicable Topics in Linear Algebra*, John Wiley and Sons, New York (1987).

J.H. Greene, *Production and Inventory Control Handbook*, 3rd edn. McGraw-Hill, New York (1997).

N.M. Grzywacz, A.L. Yuille, "Theories for the visual perception of local velocity and coherent motion". In M.S. Landy, J.A. Movshon (Eds.), *Computational models of visual processing*, pp. 231–252, Cambridge, Massachusetts: MIT Press (1991).

M. Haffner, M. Elsinga, *Urban renewal performance in complex networks Case studies in Amsterdam North and Rotterdam South*, W16 – Institutional and Organizational Change in Social Housing Organization in Europe, Int. Conference on Sustainable Urban Areas, Rotterdam (2007).

T. Hafting, M. Fyhn, S. Molden, M.-B. Moser, E.I. Moser, "Microstructure of a spatial map in the entorhinal cortex." *Nature* **436**, 801–806 (2005).

P. Haggett, R. Chorley (eds.), *Socio-Economic Models in Geography*, London, Methuen (1967).

P. Haggett, R. Chorley, *Network Analysis in Geography*, Edward Arnold, London (1969).

F.A. Haight, *Handbook of the Poisson Distribution*, Wiley, New York (1967).

J.M. Hanson, *Order and Structure in Urban Space: A Morphological History of London*, Ph.D. Thesis, University College London (1989).

G. Hatfield, "Representation and constraints: the inverse problem and the structure of visual space", *Acta Psychologica* **114**: 355–378 (2003).

A. Held, "Die Bielefeld-Verschwörung", available at http://www. bielefeldver-schwoerung.de (in German) (1994).

J. Henderson, "Economic Theory and the Cities." *Amer. Econ, Rev.* **LXIV**, 640 (1974).

M.R. Henzinger, R. Motwani, C. Silverstein, "Challenges in Web Search Engines", *Proc. of the* 18th *Int. Joint Conf. on Artificial intelligence*, 1573–1579 (2003).

L. Hermer, E.S. Spelke, "A geometric process for spatial reorientation in young children." *Nature* **370**, 57–59 (1994).

R.C. Hill, C.F. Sirmans, J.R. Knight, *Reg. Sci. Urban Econ.* **29** (1), 89–103 (1999).

B. Hillier, J. Hanson, *The Social Logic of Space* (1993, reprint, paperback edition ed.). Cambridge: Cambridge University Press (1984).

B. Hillier, R. Burdett, J. Peponis, A. Penn, "Creating life: or, does architecture determine anything?"*Architecture and Comportment / Architecture and Behavior* **3**(3), 233–250 (1987).

B. Hillier, *Space is the machine. A configurational theory of architecture*, Cambridge University Press, (1996, 1999).

B. Hillier, "A theory of the city as object: or, how spatial laws mediate the social construction of urban space", *Urban Des. Int.*, **7**, pp. 153–179 (2002).

B. Hillier, *The common language of space: a way of looking at the social, economic and environmental functioning of cities on a common basis*, Bartlett School of Graduate Studies, London (2004).

B. Hillier, "The art of place and the science of space", *World Architecture* **11**/2005 (185), Beijing, Special Issue on Space Syntax pp. 24–34 (in Chinese), pp. 96–102 (in English) (2005).

B. Hillier, *From Object- to Space-based Models of the City. Some new phenomena in the passage from model to theory*, Workshop: Evolution and Structure of Complex Systems and Networks, Zentrum fuer Interdisciplinaere Forschung, Bielefeld, February 25–29, 2008 (Germany).

R.A. Horn and C.R. Johnson, *Matrix Analysis*, Cambridge University Press, 1990 (Chap. 8).

B.D. Hughes, *Random Walks and Random Environments*, Oxford Univ. Press. (1996).

S. Iida, B. Hillier, "Network and psychological effects in urban movement", in A.G. Cohn, D.M. Mark (eds) *Proc. of Int. Conf. in Spatial Information Theory: COSIT 2005* published in *Lecture Notes in Computer Science* **3693**, 475–490, Springer-Verlag (2005).

Y. Ihara, "On discrete subgroups of the two by two projective linear group over p-adic fields", *J. Math. Soc. Japan* **18**, 219–235 (1966).

B. Jabbari, Zh. Yong, F. Hillier, "Simple random walk models for wireless terminal movements", *Vehicular Technology Conf.*, 1999 IEEE 49th **3**, 1784–1788 (Jul 1999).

M.R. Jerrum, A. Sinclair, "Approximating the Permanent". *SIAM J. Comput.* **18** (6) 1149–1178 (1989).

B. Jiang, "A space syntax approach to spatial cognition in urban environments", Position paper for NSF-funded research workshop *Cognitive Models of Dynamic Phenomena and Their Representations*, October 29–31, 1998, University of Pittsburgh, Pittsburgh, PA (1998).

B. Jiang, Ch. Claramunt, and B. Klarqvist, "An integration of space syntax into GIS for modeling urban spaces", , *International Journal of Applied Earth Observations and Geoinformation*, International Institute for Aerospace Survey and earth Sciences (ITC), Enschede, The Netherlands, **2**(3/4), pp. 161–171, Vol. **2** (3/4), 161–171 (2000).

B. Jiang, C. Claramunt, "Topological analysis of urban street networks". *Environ. Plan. B: Plan. Des.* **31**, 151–162 (2004).

I.T. Jolliffe, *Principal Component Analysis* (2-nd edition) Springer Series in Statistics (2002).

W.-S. Jung, F. Wang, H.E. Stanley, "Gravity model in the Korean highway", *Europhys. Lett. (EPL)* **81**, 48005 (6pp) doi:10.1209/0295-5075/81/48005 (2008).

M. Kac, "On the Notion of Recurrence in Discrete Stochastic Processes", *Bulletin of the American Mathematical Society* **53**, pp. 1002–1010 (1947) [Reprinted in Kac's *Probability, Number Theory, and Statistical Physics: Selected Papers*, pp. 231–239]

K.J. Kansky, *Structure of Transportation Networks: Relationships Between Network Geometry and Regional Characteristics*, Research Paper **84**, Department of Geography, University of Chicago , Chicago, IL (1963).

K. Karhunen, "Zur Spektraltheorie stochatischer Prozesse". *Ann. Acad. Sci. Fennicae* **A:1** (1944).

K. Karimi, "The spatial logic of organic cities in Iran and in the United Kingdom" in *Proc. 1st International Space Syntax Symposium*, M. Major, L. Amorim, F. Dufaux (eds), University College London, London, vol. **1**, 05.1–05.17 (1997).

R. Karp, "Reducibility Among Combinatorial Problems". *Proc. Symp. on the Complexity of Computer Computations*, Plenum Press (1972).

R.W. Kates, T.M. Parris, "Long-term trends and a sustainability transition". *Proc. Natl. Acad. Sci. USA* **100**:8062–8067 (2003).

S. M. Keane, *Stock Market Efficiency*, Oxford: Philip Allan Ltd. (1983).

S.H. Kellert, "Space perception and the fourth dimension", *Man and World* **27**: 161-180, Springer Netherlands (1994).

H. Kimon, *Religion and New Immigrants: A Grantmaking Strategy at The Pew Charitable Trusts*. Religion Program, the Pew Charitable Trusts (2001).

B. Klarqvist, 1993 A Space Syntax Glossary, *Nordisk Arkitekturforskning* **2**.

J. Kleinberg, "Authoritative sources in a hyperlinked environment". *J. ACM* **46**, 604–632 (1999).

J.J. Koenderink, A.J. van Doorn. "Affine structure from motion". *J. Opt. Soc. Am. A* **8**, 377–385 (1991).

G. Kolata, "What if they closed the 42nd Street and nobody noticed?", *The New York Times*, Dec. 25 (1990).

R.I. Kondor, J. Lafferty, "Diffusion kernels on graphs and other discrete structures". In C. Sammut and A. G. Hoffmann, editors, *Machine Learning*, Proceedings of the 19th International Conference (ICML 2002), pages 315–322. San Francisco, Morgan Kaufmann, (2002).

B.O. Koopman, "Hamiltonian Systems and Transformations in Hilbert Space". *Proc. Natl. Acad. Sci. USA* **17** 315–318 (1931).

M.J.T. Kruger, *On node and axial grid maps: distance measures and related topics.* Other. Bartlett School of Architecture and Planning, UCL, London, UK (1989).

M. Lackenby, "Heegaard splittings, the virtually Haken conjecture and Property (tau)." *Invent. Math.* **164**, 317-359 (2006).

J.-L. Lagrange, *Œuvres*, **1**, 72–79, Gauthier-Villars (1867) (in French). This remark has been given in deVerdiere 1998, p.5.

H. Ling, K. Okada "Diffusion Distance for Histogram Comparison", *Proc. IEEE Comp. Soc. Conf. Computer Vision and Pattern Recognition* (2006).

N. Linial, A. Wigderson, *Course 67659 Expander graphs and their applications*, Lecture notes available on-line on the web-page of the institute of Advanced Studies, Princeton, New Jersey (2005); also in N. Linial, *Discrete Mathematics: Expanders Graphs and Eigenvalues*, Research Channel, Video Library produced by: University of Washington (2005).

A. Lösch, *The Economics of Location*, Yale University Press, New Haven (1954).

M. Loève, *Probability Theory*, van Nostrand, New York (1955).

L. Lovász, 1993 Random Walks On Graphs: A Survey. *Bolyai Society Mathematical Studies* **2**: *Combinatorics, Paul Erdös is Eighty*, Keszthely (Hungary), p. 1–46.

L. Lovász, P. Winkler, *Mixing of Random Walks and Other Diffusions on a Graph.* Surveys in combinatorics, Stirling, pp. 119–154, London Math. Soc. Lecture Note Ser. **218**, Cambridge Univ. Press (1995).

S. Low, L. Zuniga (eds.), *The Anthropology of Space and Place: Locating Culture*, Blackwell Publishing (2003).

B. Lucas, *Which side of the road do they drive on?* http://www. brianlucas.ca/roadside/ (2005).

Q.-T. Luong, T. Viéville, "Canonic representations for the geometries of multiple projective views". In *Proceedings of the 3rd European Conference on Computer Vision*, J.-O. Eklundh (ed.) Vol. **800/801** of Lecture Notes in Computer Science, **1**, 589–599, Springer-Verlag, Berlin (1994).

U. von Luzburg, M. Belkin. O. Bousquet, "Consistency of Spectral Clustering", *Technical Report N. TR-134*, Max-Planck-Institut fuer biologische Kybernetik (2004).

K. Lynch, *The Image of the City*, The M.I.T. and Harvard University Press, Cambridge, Mass (1960).

Sh.-K. Ma, *Statistical mechanics*, World Scientific (1985).

M.C. Mackey, *Time's Arrow: The Origins of Thermodynamic Behavior*, Springer (1991).

M. Major, "Are American cities different? if so, how do they differ" in *Proc. 1st International Space Syntax Symposium* (1997). In: M. Major, L. Amorim, F. Dufaux (eds), University College London, London, **3**, 09.1–09.14 (1997).

H.A. Makse, A. Halvin., H.E. Stanley, "Modeling urban growth pattern". *Nature* **377**, 608–612 (1995).

J. Margules, C.R. Gallistel, "Heading in the rat: Determination by environmental shape." *Anim. Learn. Behav.* **16**, 404–410 (1988).

A.A. Markov, *Extension of the limit theorems of probability theory to a sum of variables connected in a chain*, reprinted in Appendix B of: R. Howard. *Dynamic Probabilistic Systems*, **1**: *Markov Chains*, John Wiley and Sons (1971).

L.E. Mays, D.L. Sparks, "Dissociation of visual and saccade-related responses in superior colliculus neurons". *Science* **230**, 1163–1165 (1980).

B.L. McNaughton, F.P. Battaglia, O. Jensen, E.I. Moser, M.-B. Moser, "Path integration and the neural basis of the *cognitive map*". *Nature Rev. Neurosci.* **7**, 663–678 (2006).

V. Medeiros, F. Holanda, "Urbis Brasiliae: investigating topological and geometrical features in Brazilian cities". In: A. Van Nes (ed.), *Proc. 5th International Space Syntax Symposium*, Delft, Faculty of Architecture, Section of Urban Renewal and Management, pp. 331–339 (2005).

M. Merleau - Ponty, *The Phenomenology of Perception*, trans. C. Smith. London: Routledge & Kegan Paul (1962).

R.K. Merton, "The Matthew effect in science." *Science* **159**, 56–63 (1968).

J.G. Miller, *Living Systems*, McGraw-Hill, New York (1978).

H. Minc, *Nonnegative matrices*, John Wiley and Sons, New York, ISBN 0-471-83966-3 (1988).

A. Möbius, *Der barycentrische Calcul*, Hildesheim, Germany (1827).

J. Morris, *Venice*, 3rd revised edition, Faber & Faber (1993).

T. Morris, *Computer Vision and Image Processing*. Palgrave Macmillan. ISBN 0-333-99451-5 (2004).

B. Nadler, S. Lafon, R.R. Coifman and I.G. Kevrekidis, "Diffusion maps, spectral clustering and reaction coordinate of dynamical systems", *Applied and Computational Harmonic Analysis: Special issue on Diffusion Maps and Wavelets*, **21**,113–127 (2006).

J. Nesetril, E. Milková, H. Nesetrilová, *Otakar Boruvka on Minimum Spanning Tree Problem (translation of the both 1926 papers, comments, history)*, CiteSeer, DMATH: Discrete Mathematics (2000).

S. Newcomb, "Note on the frequency of use of the different digits in natural numbers", *Am. J. Math.* **4**, 39–40 (1881).

M.E.J. Newman, "The Structure and Function of Complex Networks." *SIAM Revi.* **45**, 167–256 (2003).

J. Norman, H. L. MacLean, and Ch. A. Kennedy, "Comparing High and Low Residential Density: Life-Cycle Analysis of Energy Use and Greenhouse Gas Emissions." *J. Urban Plan. Dev.* **132**(1), 10–21 (2006).

J.F. Normann, J.T. Todd, "The perceptual analysis of structure from motion for rotating objects undergoing affine stretching transformations", *Percept. Psychophys.* **53**, 279–291 (1993).

J. O'Keefe, "Place units in the hippocampus of the freely moving rat." *J. Exp. Neurol.* **51**, 78–109 (1976).

J. O'Keefe, L. Nadel, *The Hippocampus as a Cognitive Map*, Clarendon, Oxford (1978).

J. O'Keefe, "Cognitive maps in infants?" *Nature* **370**, 19-20 (1994).

S. Orford, D. Dorling, R. Mitchell, M. Shaw & G. Davey-Smith, "Life and death of the people of London a historical GIS of Charles Booth's inquiry." *Health and Place* **8** (1), 25–35 (2002).

P. Ortega-Andeane, E. Jiménez-Rosas, S. Mercado-Doménech, and C. Estrada-Rodrýguez, "Space syntax as a determinant of spatial orientation perception." *Int. J. of Psychol*, **40** (1), 11–18 (2005).

E. Parzen, "On estimation of a probability density function and mode". *Ann. Math. Stat.* **33**, 1065–1076 (1962).

M. Pica Ciamarra, A. Coniglioa, "Random walk, cluster growth, and the morphology of urban conglomerations". *Physica A* **363** (2), 551–557 (2006).

L. Pietronero, E. Tosatti, V. Tosatti, A. Vespignani, "Explaining the uneven distribution of numbers in nature: the laws of Benford and Zipf". *Physica A* **293**, 297–304 (2001).

F.E. Pollick, G. Sapiro, "Constant Affine Velocity Predicts the 1/3 Power Law of Planar Motion Perception and Generation".*Vision Research*, **37** (3), 347–353 (1997).

F.E. Pollick, "The perception of motion and structure in structure-from-motion: comparison of affine and Euclidean formulations". *Vis. Res* **37** (4), 447–466 (1997).

S. Porta, P. Crucitti, V. Latora, "The Network Analysis of Urban Streets: A Dual Approach". *Physica A* **369**, 853 (2006).

A. Portes, R.G. Rumbaut. *Immigrant America: A Portrait.* (3-rd edition), University of California Press, (2006).

E. Prisner, *Graph Dynamics*, CRC Press (1995).

S.T. Rachev, *Probability Metrics and the Stability of Stochastic Models*, Wiley (1991).

N. Raford, D. R. Ragland, *Space Syntax: An Innovative Pedestrian Volume Modeling Tool for Pedestrian Safety* UC Berkeley Traffic Safety Center. Paper UCB-TSC-RR-2003-11 (December, 2003).

A. Rapoport, "A mathematical theory of motivation interaction of two individuals." *Bull. Math. Biophys.* **19**, 257–277 (1957).

C. Ratti, Space Syntax: some inconsistencies". *Environ. Plan. B: Plan. Des.* **31**, 487–499 (2004).

M. Ravallion, "Urban Poverty." *Finance Dev.* **44** (3) (2007).

W. Reed, "On the Rank-size distribution for human settlements". *J. Reg. Sci.* **42**, 1-17 (2002).

G. Rengert, "Spatial aspects of criminal behavior" in: D. Georges, K. Harris (eds.), *Crime: A Spatial Perspective*, NY, Columbia Press (1980).

L. Charbonneau, "Half of world to live in cities by end '08", the Reuters front-page on Tue Feb 26, 2008 6:51pm EST.

P. Reynolds, *Call Center Staffing*, The Call Center School Press, Lebanon, Tennessee (2003).

S.P. Richards (pseudonym of M.E. Lines see the remark of [Pietronero et al. 2001], A Number for Your Thoughts, published by S.P. Richards (1982).

K. Rosen, M, Resnick, "The size distribution of cities: an examination of the Pareto Law and primacy", *J. Urban Econ.* **8**, 165–186 (1980).

M. Rosvall, A. Trusina, P. Minnhagen, K. Sneppen, "Networks and Cities: An Information Perspective". *Phys. Rev. Lett.* **94**, 028701 (2005).

M. Rosvall, C.T. Bergstrom, "Maps of random walks on complex reveal community structure", *PNAS* **105** (4), 1118–1123 (2008).

From the Rothenburg city official site, http://www.rothenburg.de (in German) (2008).

S. Saitoh, *Theory of Reproducing Kernels and its Applications*, Longman Scientific and Technical, Harlow, UK (1988).

L. Saloff-Coste, *Lectures on Finite Markov Chains*, Ecole d'Été, Saint-Flour, Lect. Notes Math. **1664**, Springer (1997).

F. Sargolini, M. Fyhn, T. Hafting. B.L. McNaughton, M.P. Witter, M.-B. Moser, E.I. Moser, "Conjunctive representation of position, direction, and velocity in entorhinal cortex." *Science* **312** (5774), 758–762 (2006).

V.M. Savage, G.B. West, *Biological Scaling and Physiological Time: Biomedical Applications*, Springer US (2006).

S. Scellato, A. Cardillo, V. Latora and S. Porta, "The backbone of a city". *Eur. Phys. J. B* **50**, 221 (2006).

F. Semboloni, "Hierarchy, cities size distribution and Zipf's law", *Eur. Phys. J. B*, Published on line DOI: 10.1140/epjb/e2008-00203-1 (2008).

H. A. Simon, "On a class of skew distribution functions." *Biometrika* **42**, pp. 425–440 (1955).

Ya. G. Sinai, A. B. Soshnikov, *Functional Analysis and Its Applications* **32** (2), 114–131 (1998).

J.S. Shaw, R.D. Utt, (eds.): *A Guide to Smart Growth: Shattering Myths, Providing Solutions*. Heritage Foundation, Washington D.C. (2000).

J. Shi, J. Malik, "Normalized cuts and image segmentation". *IEEE Tran PAMI* **22**(8), 888–905 (2000).

G.E. Shilov, B.L. Gurevich, *Integral, Measure, and Derivative: A Unified Approach*, Richard A. Silverman (trans. from Russian), Dover Publications (1978).

A. Smola, R.I. Kondor. "Kernels and regularization on graphs". In *Learning Theory and Kernel Machines*, Springer (2003).

K. Tong Soo, "Zipf's Law for Cities: A Cross Country Investigation", *Regional Sci. Urban Economics* **35**(3), 239-263 (2002).

G. Sparr, "An algebraic-analytic method for reconstruction from image correspondences", 7th *Scandinavian conference on Image Analysis,* Aalborg, Denmark (1991).

M. Stallmann, *The 7/5 Bridges of Königsberg/ Kaliningrad*, Published on the personal web page of M. Stallmann (July 2006).

S. Sun, Ch. Zhang, Yi. Zhang, " Traffic Flow Forecasting Using a Spatio-temporal Bayesian Network Predictor", a chapter in *Artificial Neural Networks: Formal*

Models and Their Applications, pp. 273–278, Book Series Lecture Notes in Computer Science, Vol. 3697/2005, Springer (2005).

B. Tadic, "Exploring Complex Graphs by Random Walks", In P. Garrido and J. Marro (Eds.), *Modeling of complex systems: Seventh Granada Lectures*, Granada, Spain, AIP Conference Proceedings, **661**, pp. 24–26, Melville: American Institute of Physics, (2002).

M. Taniguchi, Y. Sakura, T. Uemura, "Effects of urbanization and global warming on subsurface temperature in the three mega cities in Japan". *American Geophysical Union, Fall Meeting 2003, abstract #U51A-03* (2003).

C. Tannier, D. Pumain, "Fractals in urban geography: a theoretical outline and an empirical example", *Cybergeo*, article **307**, 20 (2005).

J.S. Taube, "Head direction cells and the neurological basis for a sense of direction", *Prog. Neurobiol.* **55**, 225-256 (1998).

P. Taylor, *What Ever Happened to Those Bridges?* , Australian Mathematics Trust, University of Canberra (Math. Competitions, December 2000).

M. Tierz, E.Elizalde, *Heat Kernel-Zeta Function Relationship Coming From The Classical Moment Problem*, arXiv:math-ph/0112050 (2001).

"*Dynamics of the European urban network*", Final report realised by L.Sanders (coord.), J-M Favaro, B.Glisse, H.Mathian, D.Pumain in the framework of the European program "*Time-Geographical approaches to Emergence and Sustainable Societies*", (TIGrESS ,EVG3-2001-00024, directed by N. Winder, www.tigress.ac), March 2006, with collaborations of A.Bretagnolle, C.Buxeda, C.Didelon, F.Durand-Dasts, F.Paulus, C.Rozenblat, C.Vacchiani.

W. Thabit, *How East New York Became a Ghetto*, NYU Press (2003).

L.N. Trefethen, D. Bau, *Numerical Linear Algebra*, Society for Industrial and Applied Mathematics, ISBN 0-89871-361-7 (1997).

A. Turner, Analyzing the visual dynamics of spatial morphology. *Environment and Planning B: Planning and Design* **30**: 657–676 (2003).

A. Turner, From axial to road-centre lines: a new representation for space syntax and a new model of route choice for transport network analysis. *Environment and Planning B: Planning and Design* **34**(3):539–555 (2007).

UK Home Office, Crime Prevention College Conference, *What Really Works on Environmental Crime Prevention*, October 1998. London UK.

Population Division of the Department of Economic and Social Affairs of the United Nations Secretariat. 2007. World Population Prospects: The 2006 Revision. Dataset on CD-ROM. New York: United Nations.

US Census Bureau. *Income, Poverty, and Health Insurance Coverage in the United States* (Report P60-233) (2006).

A. Vásquez, "Exact Results for the Barabási Model of Human Dynamics." *Phys. Rev. Lett.* **95**, 248701 (2005).

A. Vázquez, J. G. Oliveira, Z. Dezsö, K. I. Goh, I. Kondor, and A. L. Barabasi. "Modeling bursts and heavy tails in human dynamics." *Phys. Rev. E* **73** 036127 (2006).

L. Vaughan, "The relationship between physical segregation and social marginalization in the urban environment." *World Architecture*, **185**, 88–96 (2005).

L. Vaughan, D. Chatford, O. Sahbaz, *Space and Exclusion: The Relationship between physical segregation. economic marginalization and poverty in the city*, Paper presented to Fifth Intern. Space Syntax Symposium, Delft, Holland (2005).

Y. Colin de Verdiére, *Spectres de Graphes*, Cours Spécialisés 4, Société Mathématique de France (1998) (in French).

D. Volchenkov, Ph. Blanchard, Random walks along the streets and canals in compact cities: Spectral analysis, dynamical modularity, information, and statistical mechanics. *Physical Review E* **75**(2) (2007).

D. Volchenkov, Ph. Blanchard, "Scaling and Universality in City Space Syntax: between Zipf and Matthew." *Physica A* **387** (10) pp. 2353–2364 (2008).

D. Volchenkov, Ph. Blanchard, "Nonlinear Diffusion through Large Complex Networks with Regular Subgraphs", *J. Stat. Phys.* **127** (4), 677–697 (2007).

J. Wagemans, A. de Troy, L. van Gool, D.H. Foster, J.R. Wood, "Minimal information to determine affine shape equivalence". *Univ. of Louven TR*, Dept. of Physiology **169** (1994).

G. Wahba, *Spline Models for Observational Data*, Vol. **59** of CBMS-NSF *Regional Conference Series in Applied Mathematics*, SIAM, Philadelphia (1990).

L. Wasserman, *All of Statistics: A Concise Course in Statistical Inference*. Springer Texts in Statistics (2005).

D.J. Watts, S.H. Strogatz, "Collective dynamics of 'small-world' networks." *Nature* **393**(6684), 440–442 (1998).

J. Weickert, *Anisotropic Diffusion in Image Processing*, ECMI Series, Teubner-Verlag, Stuttgart, Germany ISBN 3-519-02606-6 (1998).

D.M. Wilkie, "Evidence that pigeons represent Euclidean properties of space", *Journ. Exp. Psychology: Animal Behavior Process* **15**(2), 114–123 (1989).

B. Williams, *The making of Manchester Jewry 1740-1875*, Manchester UP, Manchester (1985).

L. Wirth, *The Ghetto* (edition 1988) Studies in Ethnicity, transaction Publishers, New Brunswick (USA), London (UK) (1928).

S.-J. Yang, "Exploring complex networks by walking on them", *Phys. Rev. E* **71**, 016107, (2005).

S.X. Yu, J. Shi, "Multiclass spectral clustering" in *International Conference on Computer Vision*, (2003).

H. Zha, C. Ding, M. Gu, X. He, H. Simon, *Neural Information Processing Systems* vol.**14** (NIPS 2001). 1057-1064, Vancouver, Canada. (2001).

X. Zhu, Z. Ghahramani, J. Lafferty, "Semi-supervised learning using Gaussian fields and harmonic functions". In Proc. *20th International Conf. Machine Learning*, Vol. **20**, 912 (2003).

Zipf, G.K., 1949 *Human Behavior and the Principle of Least-Effort*. Addison-Wesley.

D. Zwillinger, (Ed.). *Affine Transformations.* §4.3.2 in *CRC Standard Mathematical Tables and Formulae*. Boca Raton, FL: CRC Press, pp. 265–266 (1995).

Afterword

We have discussed the object-based, the space-based, and the time-based representations of urban environments and suggested a variety of spectral methods that can be used in order to spot the relatively isolated locations and neighborhoods, to detect urban sprawl, and to illuminate the hidden community structures in complex urban textures. The approach may be implemented for the detailed expertise of any urban pattern and the associated transport networks that may include many transportation modes.

We hope that our book will be considered an important milestone in studies seeking a quantitative theory of urban organization.

Urbanization has been the dominant demographic trend worldwide during the last half century (see Fig. 1.2). Rural to urban migration, international migration, and the reclassification or expansion of existing city boundaries have been among the major reasons for increasing urban population. The essentially fast growth of cities in the last decades urgently calls for a profound insight into the common principles stirring the structure of urban developments all over the world.

It is obvious that there is a strong positive link between national levels of human development and urbanization levels. However, even as national output is rising, the implications of rapid urban growth include increasing unemployment, lack of urban services, and overburdening of existing infrastructure that results in a decline of the quality of life for a majority of the population. Attention should be given to cities in the developing world where the accumulated urban growth is expected to double during the next 25 years.

The need could not be more urgent and the time could not be more opportune. We must act now to sustain our common future in the city.

Index

Understanding Complex Systems

Jirsa, V.K.; Kelso, J.A.S. (Eds.)
Coordination Dynamics: Issues and Trends
XIV, 272 p. 2004 [978-3-540-20323-0]

Kerner, B.S.
The Physics of Traffic:
Empirical Freeway Pattern Features,
Engineering Applications, and Theory
XXIII, 682 p. 2004 [978-3-540-20716-0]

Kleidon, A.; Lorenz, R.D. (Eds.),
Non-equilibrium Thermodynamics
and the Production of Entropy
XIX, 260 p. 2005 [978-3-540-22495-2]

Kocarev, L.; Vattay, G. (Eds.)
Complex Dynamics in Communication
Networks
X, 361 p. 2005 [978-3-540-24305-2]

McDaniel, R.R.Jr.; Driebe, D.J. (Eds.)
Uncertainty and Surprise in Complex Systems:
Questions on Working with the Unexpected
X, 200 p. 2005 [978-3-540-23773-0]

Ausloos, M.; Dirickx, M. (Eds.)
The Logistic Map and the Route to Chaos –
From the Beginnings to Modern Applications
XX, 413 p. 2006 [978-3-540-28366-9]

Kaneko, K.
Life: An Introduction to Complex Systems
Biology
XIV, 369 p. 2006 [978-3-540-32666-3]

Braha, D.; Minai, A.A.; Bar-Yam, Y. (Eds.)
Complex Engineered Systems – Science Meets
Technology
X, 384 p. 2006 [978-3-540-32831-5]

Fradkov, A.L.
Cybernetical Physics – From Control of Chaos
to Quantum Control
XII, 241 p. 2007 [978-3-540-46275-0]

Aziz-Alaoui, M.A.; Bertelle, C. (Eds.)
Emergent Properties in Natural
and Artificial Dynamical Systems
X, 280 p. 2006 [978-3-540-34822-1]

Baglio, S.; Bulsara, A. (Eds.)
Device Applications of Nonlinear Dynamics
XI, 259 p. 2006 [978-3-540-33877-2]

Jirsa, V.K.; McIntosh, A.R. (Eds.)
Handbook of Brain Connectivity
X, 528 p. 2007 [978-3-540-71462-0]

Krauskopf, B.; Osinga, H.M.;
Galan-Vioque, J. (Eds.)
Numerical Continuation Methods
for Dynamical Systems
IV, 412 p. 2007 [978-1-4020-6355-8]

Perlovsky, L.I.; Kozma, R. (Eds.)
Neurodynamics of Cognition and Consciousness
XI, 366 p. 2007 [978-3-540-73266-2]

Qudrat-Ullah, H.; Spector, J.M.; Davidsen, P. (Eds.)
Complex Decision Making – Theory and Practice
XII, 337 p. 2008 [978-3-540-73664-6]

beim Graben, P.; Zhou, C.; Thiel, M.;
Kurths, J. (Eds.)
Lectures in Supercomputational Neurosciences –
Dynamics in Complex Brain Networks
X, 378 p. 2008 [978-3-540-73158-0]

Printing: Krips bv, Meppel, The Netherlands
Binding: Stürtz, Würzburg, Germany